Flat Truth

Flat Truth

Mark Steven Hollander

Copyright © 2018 by Mark Steven Hollander.

Library of Congress Control Number:		2018905576
ISBN:	Hardcover	978-1-9845-2752-3
	Softcover	978-1-9845-2751-6
	eBook	978-1-9845-2755-4

All rights reserved. No part of this book may be reproduced or transmitted in any form or by any means, electronic or mechanical, including photocopying, recording, or by any information storage and retrieval system, without permission in writing from the copyright owner.

THE HOLY BIBLE, NEW INTERNATIONAL VERSION®, NIV® Copyright © 1973, 1978, 1984, 2011 by Biblica, Inc.® Used by permission. All rights reserved worldwide.

"Scripture taken from the NEW AMERICAN STANDARD BIBLE®, Copyright © 1960,1962,1963,1968,1971,1972,1973,1975,1977,1995 by The Lockman Foundation. Used by permission."

Any people depicted in stock imagery provided by Getty Images are models, and such images are being used for illustrative purposes only. Certain stock imagery © Getty Images.

Print information available on the last page.

Rev. date: 06/22/2018

To order additional copies of this book, contact:
Xlibris
1-888-795-4274
www.Xlibris.com
Orders@Xlibris.com

This book is dedicated to my mother who
always wanted me to become a writer.

Foreword by Donal O'Tnuthghail

MODERNISTIC VIEW

In the present modernistic worldview of today, what we call science is thought to explain most everything of the world we live in; and what it doesn't know, it is investigating with an open mind toward new discoveries because that is what science is supposed to be about. It is supposed to be based on the systematic study of the world through observations that are repeatable and measurable. At least that is what we have been taught and that is what many people assume to be true and, consequently, the basis for their belief. It is a belief based on trust—a trust we take for granted because our modernistic teachers were also taught or told the same thing. What people generally fail to realize, however, is that their trust and, consequently, their belief is misplaced. Closer examination through the real science of observation reveals that this modernistic worldview we believe to be real is completely divorced from actual reality and is, in fact, a deception—a great deception!

LOOKING AT THIS MODERNISTIC VIEW

Looking at this modernistic view has us living in a heliocentric world where the great gas giant of light is the center of our solar system orbited by "planets," of which the Earth is the third rock from the Sun. It's the basis for Einstein's "Theory of Relativity" and the understanding of gravity as the attraction of objects based on their mass. It is a theory that is further expounded by "The Big Bang Theory," where everything comes from nothing, in conjunction with "The Evolution Theory," which explains how life and everything evolved over time. All these theories and more are taught to us as if they have all been proven, and we need not question them. Of course, they are true because NASA sent the Apollo astronauts to the Moon. We all saw the rockets blast off, and we all saw Neil Armstrong's famous one giant leap for mankind. We all saw it happen live on TV. In fact, it must be true because we are constantly reminded of it through seeing the globe every day on TV and in movies. This is the way the world is, and everyone believes it. This is the supposed "enlightened" age of science and reason. In fact, science is always finding something new, and NASA is always making discoveries by finding new planets and solar systems and black holes and dark matter virtually every week. It is endlessly repeated and put in our face.

SUPPORTED BY TECHNOLOGY

Added to this view, we have technology, which supports this belief. There are thousands of satellites that make our communications work worldwide, bringing us sports, violence, and wars from all around the world. And of course, we have telescopes, which make all these things observable in the sky.

NO NEED TO QUESTION

Such is the modernistic worldview, and it is so self-evident that we need not question any of it; to do so would seem to most people, at any rate, to be totally nonsensical and to some even irrational. Why would any sane person question any of it? They are all facts. So strong are these facts too that the dogma of this modernistic view or belief in science does not allow for them to be questioned without scorn or ridicule or censor in the academic theatre.

Yet the irony of this situation is that such beliefs, where none of it can be questioned, is more akin to being a definition of a religion than to actual science. If it were really true science, then there would be no need to hide behind objectivity feigning to explain how most everything works, while not being able to explain why it works, while all these theories should be so simple and easy to prove that questioning any of them shouldn't be a problem. And therein lies the problem! If these theories and beliefs cannot be easily proven or demonstrated to be true, then believing in them isn't about facts it is about "scientism!"

Start to question any of the modernistic view and you are some sort of nut and conspiracy theorist. Question any topic—the Moon landings, satellites, telescopes, the globe—and as soon as you try to open the door to what is known (believed), you will be shut down or told that you are not qualified to say anything about the subject. Yet as soon as you start looking into any of these topics yourself, it soon becomes apparent that things are not what they seem! You only need scrap off a little of the veneer to find out it is not solid wood!

NASA AND THE MOON

Take the NASA Apollo Moon landings. Why are there questions about the photographs having multiple light sources? Why, in the videos, do the astronauts appear to be lifted on harnesses after they fall? Why are the Moon backdrops so similar to Stanley Kubrick's *2001: A Space Odyssey* movie? Why did Neil Armstrong say that space between the Earth and the Moon was a "deep black," while fellow Apollo astronaut Dr. Edgar Mitchell said that the stars were "ten times brighter"? How is it possible for there to be such discrepancies? And why has it been forty-five years since the last Moon mission? Why is current space travel restricted to "Low-Earth Orbit (LEO)? Heaven forbid that any women's group should demand that they send a woman to the Moon!

Simply looking into NASA and its space program brings more questions than answers! Why are they "blue screening" astronauts at the NASA Space Center to do "chroma key" video layering if these very same astronauts such as Tim Peake are on the International Space Station?

TECHNOLOGY

Take satellites too. How many people have ever seen one in the sky? How many people know satellites were the invention of science-fiction writer Arthur C. Clarke? Never mind that. How many people are even aware that 99 percent of our worldwide communications are done by underwater sea cable?

Take telescopes! So telescopes show us all the objects in the sky. How is it then that none of the telescopes can see the Moon during the "new moon" period each month or, for that matter, during the solar eclipse? How many people know how the lenses of telescopes actually work? Why are there concave mirrors in the lenses? Virtually nobody knows about how they work!

EVOLUTION THEORY

The "Evolution Theory" that man evolved from the ape, although the "missing link" is still missing, is currently based on the close matching of DNA between man and ape. While there is no fossil evidence to support such a link, Darwin claimed such evidence of evolution would be found. But wait a second—as soon as you look into this subject, you find that the 98 percent of DNA, which was thought to be junk DNA, is no longer considered to be junk at all! The point being that none our formerly junk DNA even matches apes or supports the evolution. This leads to the question, what scientists in their right minds decided that 98 percent of the DNA, that they didn't understand was "junk" in the first place? Maybe someone with an agenda.

THE BIG BANG THEORY

Go ask your science teacher about "The Big Bang Theory," and his response will be a wholehearted belief in it. When you point out that it is just another theory and that it comes from a Jesuit trained priest, you get a shocked look of indignation! Such teachers usually don't believe in God. Of course, it is a belief, however! According to the modernistic science, there is no God, but there just happens to be a miracle, and that miracle is that everything came from nothing, and some of it became organic and then some of that organic material became intelligent! Oh, wait! That's not the modernistic scientific view either. The current view of science states there is no "Intelligent Design (ID)" even though most everyone on the street thinks science believes there is ID! Let's just keep that a secret by not discussing it.

THE RELATIVITY THEORY

Now when we get to relativity theory and Einstein, what happens in the modernistic view starts to claim that science is too complicated for you to understand. It's Quantum physics! You have to be a scientist to explain it! While renown scientists such as Tesla called "Relativity Theory" "a beggar wrapped in purple" and Ernest Rutherford, the father of modern physics, called it "complete nonsense," this theory is never questioned. Nobody questions Einstein! Well, not anyone in the mainstream whose academic career depends on keeping quiet. There are, however, those who have stated that light does not travel at all and that photons are an invention! That light is actually a perturbation in the ether and that the medium absorbs the light, causing it to be faster in air than in water, which causing it to bend. It simply cannot be explained why light traveling through a glass of water speeds up again after it exits the water, there being no magical force to cause it to do so! Which proves light doesn't travel as photons! So if photons are not real, then Einstein's Nobel Prize based on explaining light via photons is not correct! Not only that, why are we not taught about all the experiments such as Sagnac's experiment that proves there is an ether? Because its existence disproves Einstein's theory, and this begs the question why have all the experiments that showed the Earth to be stationary have been ignored?

COPERNICUS AND THE HELIOCENTRIC OR "GLOBE" MODEL

And then there is the big one—the "Heliocentric Theory," which, as it so happens, comes from the Catholic cleric Nicolaus Copernicus in his *Of the Revolutions of the Heavenly Spheres* released over his deathbed in 1543. He too was associated with the Jesuits, but even so, this theory even comes with the caveat that "the hypothesis of the movement of the Earth is only one which is useful to explain certain phenomena, but it should not be considered as absolute truth." So the belief in this theory then becomes the truth in this modernistic worldview because of what? Because NASA claims it sent astronauts to the Moon and because we have satellites and telescopes too?

Now regarding the Jesuits, it should be pointed out that they were founded only three years earlier in 1540 just prior to this great Copernicus work. This is followed shortly thereafter by the Jesuits becoming the world's leading authority on astronomy, with many of the lunar craters named after them. In fact, today, NASA and the Jesuits run one of the greatest observatories in the world on Mount Graham called LUCIFER. I kid you not!

Do some research into the Jesuits, and you may wonder why they were formed! Napoleon Bonaparte called the Jesuits a military organization whose aim is "power in its most despotic exercise—absolute power, universal power, power to control the world by the volition of a single man." Okay, you might not like him. Take instead George Washington, who said "if the liberties of this country—the United States of America—are destroyed, it will be by the subtlety of the Roman Catholic Jesuit priests, for they are the most crafty, dangerous enemies to civil and religious liberty. They have instigated most of the wars of Europe."

So is it a coincidence that NASA and the Jesuits are so connected and control most, if not all, observatories? Let's skip over this for a bit.

If the heliocentric theory is indeed true, then everything about this theory should be easy to prove. Obviously, the Sun sets over the horizon and likewise, boats go over the horizon. We have seen the Sun set over the ocean, and boats do disappear. So there is no need to talk of things such as atmospheric refraction or the vanishing point. Or is there?

Today it is claimed that "no one believes that the Earth is flat," yet there are many, including the Mid-Atlantic Creation Research Association, which was "instrumental in revising the Kansas elementary school curriculum to remove references to evolution, earth history, and science methodology," stating "all experiments to demonstrate that the earth moves at all have failed. All seem to indicate the earth does not move at all."

SO WHAT IS THE TRUTH?

So what is the truth? As I have stated, the globe model should be easy to prove. Well, just start looking into it and see how easy it is to believe it. Because believing in it demands that you believe all the assumptions that go with it! The first is that the Earth rotates, but this rotation demands

SOLAR ECLIPSE

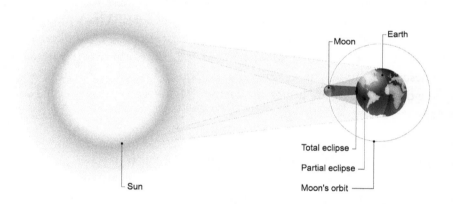

Solar eclipse.

that you believe the entire Earth atmosphere rotates with it! I never knew that before, but if the atmosphere rotates as well, why are people talking about the Coriolis effect at all? That seems a bit odd. Why is there a need to adjust for a motion of the Earth when the entire atmosphere moves?

The theory also demands, of course, that there is curvature, and that curvature works out to be six hundred feet of it from a distance of thirty miles. So how is it that sometimes on a cool day, I can see entire buildings thirty miles away across Lake Ontario from Old Fort Niagara, while at other times, the bottom parts of these very same buildings appear to be underwater? Seems when there is little or no refraction, you can see all of the buildings and when there is a lot of atmospheric refraction, you can't. So refraction causes the light to bend downward. This makes sense!

As for boats going over the horizon, it appears that they in fact disappear due to the vanishing point and atmospheric refraction, and those sunsets are just the view of the Sun refracted downward due to the very same water in the atmosphere bending the light! What's more, these observations are being made using our new technology: a Nikon P900 with eighty-three times zoom without the concave mirroring found in telescopes! But of course, these observations rely on light bending or not bending, which complicates the matter. And of course, we are told once again we are not scientists.

THE FLAW IN EARTH-ORBIT-SUN MODEL

So let's look now at the "Heliocentric" model or more descriptively "The Earth-Orbit-The Sun model. Surely the geometry of this model should easily prove we live on a globe. I mean, come on, how can the geometry of this model not be correct?

The simple premise in the modernistic model view is that the Earth travels around the Sun and at four points during the year we have the "Summer Solstice," the "Autumn Equinox," the "Winter Solstice," and the "Spring equinox." Now as we all should know by our education, the Earth spins on its axis, which is always tilted at a 23.4-degree angle pointing to the north star called Polaris. And no matter what time of year, the Earth is always pointing (tilted) toward this star. However, the Earth axis is NOT always tilted toward the Sun! (Something conveniently ignored!)

THE SUMMER SOLSTICE

At the midpoint of the summer occurs the Summer Solstice, where the tilt of the Earth is at a maximum 23.4 degrees toward both Polaris and the rays of the Sun. What this means is that in the Northern Hemisphere, as viewed from the Sun, the lines of latitude of the Earth curve downward, toward the Ecliptic as the Earth rotates from sunrise until noon before curving back upward away from the Sun toward sunset. This curve of the latitudes toward the plane of the Ecliptic is what causes the Sun to delineate in the sky toward the equator, forming a southward arc. To explain this further, take an observer standing on the Tropic of Cancer. In the morning, the observer will see the sunrise very much to the north of the tropic. As the Sun rises in the sky, it will arc toward the south until noon from whence it will then arc back toward the north before sunset.

In the geometric model, the path the Sun makes in the sky arcing southward corresponds to the movement of the observer as the latitude moves downward on the rotating Earth until noon and then back in the opposite direction toward sunset and can be expressed as a general principle of this model as "the arc of the Sun follows the latitude of the observer." This observation is confirmed in the Summer Solstice, which has a southern arc of the Sun in the sky because of the tilt.

THE AUTUMN EQUINOX

Now as the Earth continues its orbit around the Sun, the tilt toward the rays of the Sun starts to slowly diminish until at the Autumn Equinox. At this time there is no tilt of the Earth axis pointing toward the Sun, albeit, while the Earth is still pointing toward Polaris. What this means is all the latitudes on Earth are horizontal to the rays of the Sun. This horizontal position of all the latitudes is like a level "spin top" and is supposed to be what causes the day to equal night as all the longitudes on Earth get sunrise and sunset at the same time. (At least we are told they do, but that is another story for another time!) In turn, what this means to the observer on the Tropic of Cancer is that as the Earth rotates, the Sun should rise due east and traverse horizontally along the same latitude of the observer before setting due west. This follows the general principle shown by the Summer Solstice, "the arc of the Sun follows the latitude of the observer." But wait a moment—the Sun doesn't follow the observer during the equinox because it doesn't traverse along the latitude of the observer on the Tropic of Cancer during the equinox. Nor does it do that at any latitude on Earth, except at the equator. The Sun still arcs southward in the Northern Hemisphere at this time, but this is geometrically impossible on the globe model, and this is not complicated!

Not only is this flaw very evident, but the flaw also gets worse as the Earth orbits toward the Winter Solstice! But I will stop here and say only this. The equinox in the globe model is in fact the crossover point by which the lateral arc of the Sun in the sky moves from the southward arc we see in the summer in the Northern Hemisphere to what *should be* a *neutral* arc at the equinoxes changing to a *northern* arc between the equinoxes, *and this globe model geometry is not what we observe in the real world!*

THE SO-CALLED GREAT MEN OF SCIENCE

Not only does the southern arc of the Sun in the sky in the Northern Hemisphere at the equinox prove that the globe model is flawed. Again, this geometry is not rocket science, and it means something far more significant! It means that the so-called great men of science, who have been promoted to be of such high esteem such as Copernicus, Galileo, Newton, and Einstein, would or should all have known that this model is flawed! It means that what we are being told about these "great" men are nonsense! It means that Tesla, who stated that "the earth is an enclosed realm," knew the truth. It means the ancient cultures knew the truth, and it means that those professing the spinning ball Earth are in fact clearly deceiving you!

SO WHY IS IT SO EASY TO BELIEVE THE LIE?

It would appear that the bigger the lie, the easier it is to believe. Who would put forth such lies? Yet it is also true that by constant repetition of putting forth so-called statements of fact and theories as true that the analytical mind has a tendency to believe anything regardless of whether it is true or false. It only requires the mind to accept the premise that something is true.

As Adolf Hitler stated in *Mein Kampf*: "In the big lie there is always a certain force of credibility; because the broad masses . . . more readily fall victims to the big lie than the small lie, since they themselves often tell small lies in little matters but would be ashamed to resort to large scale falsehoods. It would never come into their heads to fabricate colossal untruths, and they would not believe that others could have the impudence to distort the truth so infamously."

This means that the tendency to believe the lie stems more on repetition than on real facts, and there is a strong tendency on those who are educated with these facts to support them true or false. What it takes to discern the truth is the ability of someone to stop accepting facts without the science of observation and reason. Too many people think they know the truth when they have actually based their reasoning on assumptions based on beliefs! This is not science but scientism.

If you want to know the truth, you have to rid yourself of all the brainwashing or indoctrination that you have been told and start looking at things with a clear new perspective. Call it an awaking. Some might call it being "reborn." Whatever you call it, when you start rubbing on the veneer, you start discovering there is no real wood! We don't live on a Jesuit spinning ball Earth, and the truth of that is real science and observation and not accepting the modernistic view of the world you were lied to about. There is no curvature to the Earth. The earth is flat!

Donal O'Tnuthghail, January 21, 2018

Hoodwinked: Into the Age of Insanity

Part 1

Past Deception

SIN IN THE GARDEN

In Genesis, humans are pictured as the special creation of God (Genesis 2:7) placed in a garden created by God (Genesis 2:8–15), and three very critically important features are crucial for understanding the human role in the garden:

1. Adam was put in the garden to cultivate and keep it. God in the narration for the vocation for man's fulfilment.
2. The first people were granted great freedom and discretion in the garden. This freedom permitted them to take from the goodness of God's creation (Genesis 2:16). The freedom they had as well as the discretion was very limited.
3. The tree of knowledge: With the tempting fruit that was appetizing was from the unveiling of God's discernment that carried fruit of good as well as evil. God prohibited the taking of any fruit because he is all knowing, and any disobedience would prohibit man's ascension that God had in store for him. Many scholars have pointed out that these three features belong uniquely to humans. Each person faces his or her own vocation, freedom, and prohibition. I am of the belief that full humanity is experienced only when all three of these are maintained. The knowledge of good and evil would make humans godlike in some ways (Genesis 3:5

and 22). Biblical scholars understand these three virtues as holding ultimate knowledge and needs to be experienced to obtain freedom of choice. Furthermore, man has to learn from life's bad choices, to learn from that bad experience, and to grow from it physically or in the flesh and also spiritually, which is your transfer as one, the heart interfacing to the mind. To be able to reflect on bad choices and learn from them as a seed of growth of discernment and true knowledge of God, our creator, and savior of mankind. This lesson resonates as feeding your very soul for growth toward God. The failed realization of the opposite will ultimately disconnect man and the living spirit of God. The serpent in Genesis is identified only as a creature, but theological experts, upon reflection, has definitely connected the serpent as the instrument of Lucifer and is thus legitimately cursed and pictured as the enemy of woman's seed (Genesis 3:14–15). The serpent was more clever than any of the wild animals that our creator made. The serpent said to the woman called Eve, "Did God really say you must not eat fruit from any tree in the garden?"

Eve said to the serpent, "We may eat fruit from the trees in the garden. But God did say you must not eat the fruit from the tree in the middle of the garden. Do not even touch it. If you do, you will die."

"You will certainly not die," the serpent said to the woman. "God knows that when you eat fruit from that tree, you will know things—such as technology beliefs of today from lies—you have never known before. Like God, you will be able to tell the difference between good and evil."

Eve ate from the forbidden fruit, and it was very beautiful to look at before digesting it, and she saw that it made her very wise and shared it with Adam, who ate the fruit and knew things that they never knew before. They started using fig trees to cover their nakedness. Then Adam and the woman called Eve, upon hearing the Lord God come through the brush, immediately hid from God, knowing already why God was summoning them.

The Lord God asked, "Have you eaten fruit from the tree I commanded you not to eat from?"

The man called Adam answered, "I was afraid because I was naked, and it is the fault of the woman that you put here with me. She gave me some fruit from the tree, and I ate it."

Then the Lord God said to the woman, "What have you done?"

The woman said, "The serpent tricked me. That's why I ate the fruit."

The Lord God spoke to the serpent, and said, "You are set apart from all livestock and all wild animals. I am putting a curse on you. You will crawl on your belly. You will eat dust all the days of your life. I will make you and the woman hate each other. Your children and her children will be enemies. Her son will crush your head, and you will bite his heel."

God said to Eve, "I will increase your pain when you give birth. You will be in great pain when you have children. You will long for your husband, and he will rule over you."

The Lord God said to Adam, "You listened to your wife's suggestion. You ate fruit from the tree I warned you about. I said you must not eat its fruit. So I am putting a curse on the ground because of what you did. All the days of your life, you will have to work hard. It will be painful for you to get food from the ground. You will eat plants from the field even though the ground produces thorns and prickly weeds. You will have to work hard and sweat a lot to produce the food you eat. You were made out of the ground. You will return to it when you die. You are dust, and you will return to dust." Before the Lord left Adam and Eve, he made clothes out of animal skins for them to wear. God said, "Just like one of us, the man can now tell the difference between good and evil. He must not be allowed to reach out and pick fruit from the tree of life and eat it. If he does, he will live forever." So the Lord God drove the man out of the Garden of Eden. He sent the man to farm the ground he had been made from. The Lord God drove him out and then placed angels on the east side of the garden and placed a flaming sword that flashed back and forth. The angels and the sword guarded the way to the tree of life.

Later on, when the revolt in heaven developed, and the fallen angels from heaven came to pervert the seed of God's ability to come back as a man. He so flooded the world, and I mean the whole landmass was overridden by the *great flood*. In some aspects of trying to determine the exact location of the Garden of Eden and some estimates, it's believed to be in the middle of the Persian Gulf in about 1,200 foot under the sea.

Part 2

The Two Hundred Fallen Angels From The Big Revolt And War In The Third Heaven

The sons of God were and are the Nephilim and descended through an inter-dimensional portal right above where Mount Hermon stands today near the Lebanese border with Syria, and you can take in its view from the Golan Heights in Israel. What people do not understand about what had happened was that these two hundred fallen angels were expelled from heaven and to enact their prideful and corruptible revenge on God, the creator, was to come back to Earth and to try to wipe out the very genealogy of any purity of godly genes and to also prevent God from being able to come in the flesh. What isn't reported in common everyday knowledge that is hard to comprehend and very little understanding about if this reality that occurred 3,500 years ago, and the prophecy that has evolved from fiction, and the reasons vary why the book of Enoch was taken from the Bible, but what if this is very real? Enoch is the only human being who never died. God took him! Enoch was taken up to heaven and shown heaven and all the wonders of the universe and the reason why the book of Enoch and the book of Eve were eliminated by the heads of the Vatican and the profiteers of hiding the true nature of who these profiteers were, and it is of the domain of the satanic agenda that has been in place since the days of the very first Freemason, right around 500 BC, and Pythagoras is revered by all these neophytes who are enemies of truth and of humanity. What Enoch saw is probably a factual true occurrence. It is not just a coincidence, yet it is suppressed from the mainstream. Come to think of it, mostly anything of purity that is of a factual occurrence from our past or an event that has already been prophesied as coming to pass or an event that maybe subjective or interpretive but seems destined to happen. We live in the times when the meaning of the "return to the days of Noah." We could possibly be in the matrix of the end times as is carefully been foretold in the book of Revelations.

Some of the North American Native Americans such as the Apache Indian tribal reservation has recently reported in the last two years near Mount Graham, several people are seeing the sky open up in a way that can be classified as anything but natural. The Native American Apache elders are having a very difficult time trying to explain what they are witnessing but seem to believe and are convinced that there is a stargate above Mount Graham. It could be the stargate that brings the interdimensional entities who are posing as the grey aliens, but the Nephilim is the real origin of these fake cosmology creatures who claim to live in other galaxies and other planets that are several gazillion light-years away from Earth. This is part of the master plan from the greatest deceiver of them all: Satan. What if we are to remain asleep during the massive feast of lies for your conscience as well as accidentally losing your very soul because you remained asleep? I honestly do not believe God punishes people for being ignorant. I believe that our system of where we are in humanity has transcended insanity to us to become normal, and the artificiality and the evolving technology that has arrived to capture us into their world of fantasy and disconnection of having any spirit nature about oneself and to create a new generation: the millennials. What is going on around the Roman Catholic-Vatican observatory near Mount Graham? Why did the Vatican fight in Federal court with NASA to build a first-class observatory with some of the finest telescopes and an unusual mountain above the Arizona desert? The unsuspecting court battle did rage a little bit, and the fact that the governor of Arizona even got involved to try to come up with a resolution between the main two power brokers of deception was quite unusual to say the least. After all, the two power brokers of the heliocentric wizardry and the ball-Earth deceivers are securing and protecting the "biggest lie in mankind's history," and they hide the very shape of our world in order to maintain power on it!

The fact that the binocular telescope inside the observatory is called LUCIFER is quite astonishing to anyone trying to understand why anyone would name a telescope after the devil? That it supposedly can also provide heat signatures that is built into this futuristic technology is quite an accomplishment to the priests within the ranks of the Catholic Church and are still today some of the world's very best astronomers. The secret agenda with the global elite heliocentric model is to teach us and stay on course with psychotic science such as Darwinism and, of course, the continued effort to hide any existence of God, the creator and the intelligent designer, who always suggested that we were in a stationary Earth, and we are ultimately the center of the universe, which provides the notion that our life and the lives that we lead are so much more important than what we are taught and continue to be taught, especially the programming poison of our young lads for the next generation, the children. The hidden archaeology in the world is a little easier with the fact that most of the giant human skeletal remains were in the late 1800s and into the early 1900s. Fact is most of the North American Native American tribes and the remaining safeguarded cultures and traditions have a deep and mostly very violent clashes with the Nephilim giants, and we will try to revisit some of that history that was at least two hundred years before the Europeans landed on Plymouth Rock to claim what was then called the new world.

Zecharia Sitchin, who wrote extensively about the Annanaki race of giants who inhabited the Earth thousands of years ago, got it very wrong. He wrote about these beings who were eight feet to over twelve feet tall in height as being an extraterrestrial race who came down and genetically fused with a woman and created giant human beings as they were, but these beings were involved in a war in the third heavenly realm and were part of the original two hundred fallen angels who not only perverted the grounds of Earth and mankind but also all the animals who occupied our world at that time. Sitchin is a Freemason paid shill who proposed the agenda of extraterrestrial beings when, in fact, these are inter-dimensional beings of God in Heaven. The mistaken identity of these giant human beings who perverted all animals on Earth but, more importantly, the very seed of God's creation: man. It's in Genesis 6:4:

> There were giants on the Earth in those days, and also afterward when the sons of God came in to the daughters

of men and they bore children to them. Those were the mighty men who were of old men of renown. (KJV)

It continues in verse 5: "Then the Lord saw that the wickedness of man was great in the earth, and that every intent of the thoughts of his heart was only 'evil' continually." The Nephilim were the fallen angels that came down, and the offspring who at least half the time were killing the women because the babies were of enormous girth and size, the giants of the days of Noah, essentially were Lucifer's forces to pervert the very deed of God and man so that everything and everyone would have absolutely no purity left, in other words, to prevent God from coming into this world to save man from this perversion from the devils of this world. The devil even tried to lure Jesus on Mount Olive to give up his sinless task to save man by offering Jesus all the riches and fame that his world can offer. Thankfully, Jesus did not submit to Lucifer's offer, and now we are as we were yesterday, and Satan and the satanic agenda are the Luciferian Baal worshipping thirty-third Freemason Illuminati laced with world domination through the money that propelled the lies of our educational institutions worldwide. Sitchin contends that the word "Nephilim" means "those who came down from above" who descended to Earth. They were the two hundred maggots who descended over Mount Hermon near Lebanon, and this is chronicled extensively in the book of Enoch. It's no accident that Enoch was taken out of the Bible because these are the skeletons in their closet. The real identity of the world's ruling elite and their ancestors Sitchin also believed that these giant Annanaki came from above "people of the fiery rockets," so there is the twelfth planet Nibiru that the space warriors of dulled conscious-less state who believe in Pinocchio, or I do mean, space.

The translations from Sitchin, essentially was of a brilliant mind that clearly submitted to the cosmic fairy-tale agenda to believe that aliens are coming to take over our world from an imaginary thousands of phony light-years away. In reality, this kind of garbage was written thousands of years before Zecharia Sitchin was even a thought! Sitchin claimed the Nephilim were ancient astronauts, and as such, it is not hard to overestimate this very notion that the beginning of his thesis is incorrect, which makes his overall thesis an absolute fairy tale not based on reality. His work is not in step with the old Hebrew Bible, and his conclusions were erroneously misidentified the word Nephilim and where their origin actually came from. Sitchin assumes Nephilim comes from the Hebrew word *naphal*, which usually means "to fall." He then forces the meaning "to come down" onto the word, creating his "to come down from above" translation, which had no merit at all. In the form we find it in the Hebrew Bible, if the word "Nephilim" came from Hebrew *naphal*. It would not be spelled as we find it. The form Nephilim cannot mean "fallen ones" because the spelling would be "nephulim." Likewise, Nephilim does not mean "those who fall" or "those who fall away" (that would be *nophelim*). The only way in Hebrew to get Nephilim from *naphal* by the rules of Hebrew morphology (word formation) would be to presume a noun spelled *naphil* and then pluralize it. I say "presume" since this noun does not exist in biblical Hebrew—unless one counts Genesis 6:4 and Numbers 13:33, the two occurrences of Nephilim, but that would be assuming or guessing, what one was trying to prove! However, the Aramaic noun for naphil (a) does exist. It means "giant," making it easy to see why the Septuagint (ancient Greek translation of the Hebrew Bible) translated Nephilim as *gigantes* (giant).

Sitchin wants to assume that Nephilim means "those who *came down* from heaven" so he can confuse the origin of their intentions and convince millions that these were ancient astronauts, which is essentially ridiculous and not based on any factual events. One of the amazing errors in his work is that he didn't decipher the differences between Hebrew and Aramaic texts, and they use the same alphabet. In Sitchin's book *Stairway to Heaven*, he omitted the description of the word "watchers" from the Dead Sea Scroll called the Genesis Apocryphon, and it clearly suggest that he did not know the Qumran text was written in Aramaic, not Hebrew, several inconsistencies that were assumptions based on a nefarious agenda. As brilliant as he was, he intentionally was writing a fairy tale for the ilk of propaganda TV shows and talk circuits. The book *Chariots of the Gods* and the popularity also became a blockbuster Hollywood movie, and he was in the very popular "ancient aliens," which, before my own awakening, was one of my favorites.

PART 1

Subject: PART 1: THE NEPHILIM, "The Fallen Watchers" AND THEIR OFFSPRING THE Rapharaim- and some of the North American Indians, most of their cultures had battle with Nephilim and

Rapharaim Giants of old and of RENOWN, 100 to 250 years before these tribes did battle with the U.S. Calvary!"

PART 1

The giants from the Bible have been hidden and is forbidden in the modernized science of archeology not to do any investigative work into the suppressed field of the "giants of renown." I will chronicle some finds throughout the United States in reference to digging up giant skeletons who did almost always turn it over to the government, and the Smithsonian Institution did a remarkable job of making any and all finds disappear and never to be seen again. It is a brilliant cover-up that just validates Scripture and what truth ultimately can be substantially found in the word of God. The newer institutes, the National Geographic and the Discovery Channel, are every bit as Illuminati-thirty-third Freemason controlled as the Smithsonian in that they are just as untruthful to where our history is. Our true reality is quite different from what these government-controlled propaganda and the cosmology and Hollywood-NASA-Disney fake reality continue to push through the mainstream as well as the continued poison of teaching this nonsense to our children. Everyone needs to know the truth, though nefarious and atheist teaching is predominant thinking to what truth really is. Below are some finds dug up throughout the United States.

In 1879, the *Indianapolis News* reported that a nine-foot, eight-inch skeleton was excavated from a large mound near Brewersville, Indiana. It was known as the giant from Brewers Cave.

In 1891, workmen excavating near Crittenden, Arizona, discovered a huge sarcophagus about ten feet below the surface, which contained the mummy of a twelve-foot tall giant with six digits on each limb.

The *Chicago Record* reported on October 24, 1895, that a mound near Toledo, Ohio, was found to hold at least twenty skeletons and teeth, seated and facing East with jaws twice as large as any present-day people. Beside the sitting dead giants were very large bowls with curiously wrought hieroglyphic figures, and my guess is it was ancient Egyptian art because this is the origin of these creatures, if I may.

April 5, 1909, the discovery that was so amazing that complete and suppressed information about this was immediately hidden from the time that the Egyptian tomb was excavated by the Smithsonian Institution that took place in Arizona. A lengthy front-page story was written by the *Phoenix Gazette* of a rock-cut vault by an expedition led by Professor S.A. Jordan of the Smithsonian.

The Smithsonian claims to have no knowledge of this discovery even though the local paper ran it as a front-page story. The area around Ninety-Four Creek and Trinity Creek had areas with names such as the Tower of Set, Tower of Ra, Horus Temple, Osiris Temple, and the Isis Temple. Near the Haunted Canyon were the names Cheops Pyramid, the Buddha Cloister, Buddha Temple, Manu Temple, and Shiva Temple. Was there any relationship between these places and the alleged Egyptian discoveries in the Grand Canyon? The authorities from the Grand Canyon do confirm that these places are definitely off-limits to hikers and tourists, and they use the reason of the dangerous caves that exist at the bottom of the Grand Canyon. No one is allowed into this large and beautiful area that transcends the power of a forbidden zone with the power of the Colorado River running into the ragged walls of the forbidden zone in the Grand Canyon with the funny hieroglyphics and the Hindu names that adorn the lower walls of this majestic landscape with an aura of splendid beauty matched by almost nothing. We can only conclude that this was the area where the vaults were located. I believe the 125-year cover-up from the people who created a Smithsonian gate right inside the halls of our most hallowed archaeological institution has been actively involved in suppressing evidence or ancient giants and other oddball artifacts, copper, pearls, gold, silver, and evidence that tends to disprove the official party line and advanced American cultures, ancient voyages, various cultures throughout North America, and suppression from any truth that proves theology, geocentric, and anything that would provide an anti-litmus of accuracy from the word of God.

I came across an article from Astral Internet after a "giant" human femur bone that was dug up in Ohio in 2011 by the American Association for Alternative Archaeology that was accompanied by a picture of the archaeologist standing next to the gigantic bone reaching up to his shoulders. If the man was average in height, then the "femur" bone measured at least five feet from the ground to his shoulder level. The find was so revealing that the AAAA sued the Smithsonian Institute and brought the matter to the U.S. Supreme Court. A U.S. Supreme Court ruling forced the Smithsonian Institution to release classified papers dating from the early 1900s that proves the organization was involved in a major historical cover-up of evidence showing giant human remains in the tens of thousands that had been uncovered all across America and were ordered to be destroyed by high-level administrators to protect the fantasy of evolution that was mainstream chronology of human evolution taught in all schools, including the most esteemed universities in the world. The allegations stemming from the American Institution of Alternative Archeology (AIAA) that the Smithsonian Institution had destroyed thousands of giant human remains during the early 1900s was not taken lightly by the Smithsonian who responded by suing the organization for defamation and trying to damage the reputation of the 168-year-old institution.

During the court case, new elements were brought to light as several Smithsonian whistle-blowers admitted to the existence of documents that allegedly proved the destruction of tens of thousands of human skeletons reaching between six feet and twelve feet in height, a reality mainstream archeology cannot admit to for different reasons, claims AAAA spokesman James Churward:

> There has been a major cover up by western archaeological institutions since the early 1900's to make us believe that America was first colonized by Asian peoples migrating through the Bering Strait 15,000 years ago, when in fact, there are hundreds of thousands of burial mounds all over America which the Natives claim were there a long time before them, and that show traces of a highly developed civilization complex use of metal alloys and where giant human skeletons remains are frequently found but still go unreported in the media and news outlets.

A turning point of the court case was when a 1.3-meter long human femur bone was shown as evidence to the court to prove the existence of such giant human bones. The evidence came as a blow to the Smithsonian lawyers, as the bone had been stolen from the Smithsonian by one of their high-level curators in the mid-1930s who had kept the bone all his life and which he had admitted on his deathbed in writing of the undercover operations of the Smithsonian.

"It is a terrible thing that is being done to the American people," he wrote in the letter. "We are hiding the truth about the forefathers of humanity, our ancestors, the giants who roamed the earth as recalled in the Bible and ancient texts of the world."

The US Supreme Court has since forced the Smithsonian Institution to publicly release classified information about anything related to the "destruction of evidence pertaining to the mound builder culture" and to the elements "relative to human skeletons of greater height than usual,", a ruling the AIAA is extremely enthused about.

"The public release of these documents will help archeologists and historians to reevaluate current theories about human evolution and help us greater our understanding of the mound builder culture in America and around the world," said Hans Guttenberg. "Finally, after over a century of lies, the truth about our giant ancestors shall be revealed to the world." He was visibly satisfied by the court ruling.

The documents are scheduled to be released in 2015, and the operation will be coordinated by an independent scientific organization to assure political neutrality. The date of the Supreme Court ruling was on December 3, 2014.

Serpent Mound, Ohio

The great Serpent Mound is a 1,370-foot prehistoric mound near Peebles that was researched by Ross Hamilton, and he wrote about many of its mysteries of the giants he discovered. Some radiocarbon analysis dates to about 321 BC, which puts it into the Adena civilization, who obviously were also present way before the Indians, whom we were taught were the only residents in the new world before the Western psychotic science that we were indoctrinated into believing. Researcher Jeffrey Wilson discovered a giant here that was cut off at the knees and it was still seven feet tall. The estimated height of the giant was about eight feet in height.

The Red-Haired Giants of Lovelock Caves, Nevada,1911

The amazing finds at Lovelock Cave, about seventy-nine miles east of Reno, Nevada, was a guano-filled cave on the shores of Lake Lahonton, in an area called the Great Basin and was originally found in 1911.These finds were discarded countless times by Smithsonian's archeologists, but the related finds at Spirit Cave brought credibility to the finds at Lovelock and is now considered an ascertained fact. Pictures of human skulls at the Humboldt Museum in Winnemucca, Nevada, sort of put to rest the story of the Paiute Indians who fought a vicious war with the giants known as the Si-Te-Cah. These red-haired tribe of cannibals who lived near the Paiutes often harassed them with continuous

Cresap Mound, West Virginia

Dr. Donald Dragoo from the Carnegie Museum, in 1959, unearthed a seven-foot, two-inch skeleton during this excavation in Northern West Virginia. "The individual was of large proportions. When measured in the tomb his length was approximately 7.04 feet. All of long bones were quite heavy."

The seven-foot, two-inch skeleton was partially burnt, which indicates maybe a brush with some of the Native Americans who have chronicled the viciousness of the wars all over the United States with the giants.

Jasper, Ohio, the nine-foot giants found in 1886

Many places in southern Ohio has many recollections and excavations of giants found, whereas the evidence has been left, especially the Serpent Mounds and many others. The farm was owned by Mr. William Bush, and another one was found on Mr. Matthew Mark's farm as reported by the *Stevens Point Daily Journal* on May 1, 1886. At the Mark bank, dozens of human skeletons have been exhumed since the bank was first opened, according to some of the locals. At least two of the skeletons measured over nine feet in height. Amazingly enough, all the skeletons were buried standing in an erect position and were comparatively well preserved. The even more incredible detail of these giants was the almost perfect state of preservation in which their teeth were found to be. Not a decayed tooth has been discovered, and this would seem to indicate that these people naturally had excellent teeth or some extraordinarily manner of preserving them.

Crowville, Louisiana, twelve feet in average height, 1914

Skeletons that averaged twelve feet in height found by workmen working on a drainage project were found in Crowville, Louisiana, and reported in the *Monroe County Mail* on June 18, 1914. Several of the skeletons were lying in various positions that indicated that they were killed in a prehistoric fight. They were covered with alluvial deposits due to the flooding from the Mississippi River. There were no weapons or artifacts found, and the Titans evidently struggled with wooden clubs. The skulls were in excellent condition, and some of the jawbones are large enough to surround a baby's body.

Mounds in Iowa in 1897

On November 18, 1897, in the Iowan newspaper the *Worthington Advance*, John Wesley Powell from the Smithsonian Institute dug up a human skeleton of a seven-foot, six-inch giant. Unfortunately for the archeologists who were on sight when the excavation was almost completed, the bones were crumbled to dust when it was exposed to the open air.

Steelville, Missouri, 1933

An eight-foot giant skeleton was found by a young boy who was looking for arrowheads and came across this giant near Steelville, Missouri, on June 11,1933. Researcher James Clary went on site and had this to say, based on the newspaper account:

> An ancient Ozark Giant dug up near Steelville: Strange discovery made by a boy looking for arrowheads, gives this Missouri Town an absorbing mystery to ponder. (*The Steelville Ledger*, June 11, 1933)

Catalina Island, California

In 1919, the Channel Islands off the coast of California have turned up several gigantic human skeletons, including one that measured nine feet, two inches, and several others that measured over eight feet tall. The story is intriguing because the archeologist was Ralph Glidden, who also had an unusual museum. One of the eight-foot skeletons were dressed in full armor that apparently was his cavalry war-monger attire.

Mr. Glidden was an amateur archeologist, but based on his incredible findings of 3,781 skeletons that he successfully excavated between 1919 and 1930, he was anything but an amateur. He went on to work for the Heye Foundation and had this to say:

> A skeleton of a young girl, evidently of high rank, within a large funeral urn, was surrounded by those of 64 children, and in various parts of the island, more than 3,000 other skeletons were found, practically all the males averaging around 7 feet in height, with one being 7 feet 8 inches from the top of his head to his ankle, and another being 9 feet 2 inches tall.

Winona, Minnesota, Giant: ten feet, three inches; nine-foot skeleton found in Dresbach, 1883

Many Indian mounds with lots of relics were found when men were digging in Mineral Bluff and a ten-foot, three-inch skeleton was unearthed. It apparently was killed because a copper hatchet was found through the skull along with an enormous arrowhead that was nine inches long, which ended this giant human being's life. Not too far away in Dresbach, a nine-foot skeleton was uncovered while some men were digging a trench. These were all taken from old Indian mounds, which may have been the reason why the giants were killed, and the sheer size, form, and structure would lead paleontologists to believe that they belonged to a race that were inhabitants prior to the Native Indian tribe that was present in the area. In many of the mounds, the finds were absolutely astounding even by today's standards. The copper hatchets, chisels, and different makes of tomahawks and other weapons of warfare verified that these ancient giants used an unknown process for hardening copper that cannot be duplicated today.

Eagle River, Iowa: seven-foot, seven-inches (double rows of teeth called dentitions), April 1900

In the *Journal Tribune*, as reported in Williamsburg, Iowa, was a well-preserved athletically looking "giant" of enormous body features with a very large and thick skull that was over a quarter of an inch and double upper and lower rows of perfectly preserved teeth. The femurs were very long, showing a giant in stature. Dr. N.C. Morse examined the skeleton and deemed this monster to have had athletic prowess that would be impossible to match in any era.

Beaver Lake, Ozark Caves, Arkansas, 1913

Victor Schoffelmeyer, a reporter from the *New Age Magazine* (volume 18, 1913), at a depth of only three feet, found the remains of a ten-foot giant along with several other giants, some having severe cranial deformities. During the filming of *Search for the Lost Giants*, the site of the cave was investigated by several other nationally esteemed archeologists from—you guessed it—the Smithsonian. The caves were flooded with the damming and creation of Beaver Lake that occurred between 1960 and 1966. Scuba divers, when looking for any last remnants under water at the site of the giant finds, were astonished to find a seventy-foot wall showing a likely area of human habitation, but no other bones could be found, especially in sixty-feet-deep of water, which more than likely dissipated any other worthy items that could be valuable for use of archeological means. A man named Dad Riggins spent much of his time digging the ashes, which formed some of the other floors from these Ozark Caves. After digging more than three feet, he came across an almost perfect skull of gigantic size, which was very different in some particulars from a modern specimen. When partly joined, the largest skeleton was almost ten feet tall. Riggins showed hieroglyphics covering the Palisades thought to be thousands of years old.

The San Diego Giant, 1895, and Spiro Mounds, Oklahoma

The discovery of a "giant mummy" was reported in a California newspaper as being a nine-foot Indian. The San Diego giant was purchased by the Smithsonian Institute for $500, and later in 1908, the institution claimed that it was a hoax, which, after being examined by several archeological experts and then coming out thirteen years later, was suggested as made of gelatin. At this time was a fight by the Illuminati-globe-deception of Darwinism, which became a secret narrative at this time to make the bones disappear or provide an official statement to discredit the very find by suggesting a hoax. Another similar mummified giant was excavated at Spiro Mounds in Oklahoma, which measured eight feet, five inches and was on display at the museum for several months before mysteriously disappearing.

The R Miamisburg, Montgomery County, Ohio

Another Mound in Montgomery County, Ohio, that was over seven feet tall, a height of a seven-story building and is the largest conical burial in Ohio, even larger than the Serpent Mound, and is an unbelievable 877 feet in circumference. This site was investigated recently in September 2012 because other giant skeletal remains were found at this site just five years ago. The Miamisburg Mound was excavated further, and the remains of a "jawbone" that can fit over a whole normal-size human head and was such an unusual find. Two of the researchers left the project after a few digs, but the dig became a national sensation and was reported and then omitted from all major media news stream for quite obvious reasons as it has since the early 1900s. The Middletown Signal on January 17, 1899, did report with the front-page headline: "Bones of Prehistoric Giant Found Near Miamisburg."

> The skeleton of a giant found near Miamisburg is the cause of much discussion not only among the curious and illiterate but among the learned scientists of the world. The body of a man more gigantic than any ever recorded in human history has been found in the Miami Valley, in Ohio. The skeleton it is calculated must have belonged to a man 8 feet 1.5 inches in height.

Prof. Thomas Wilson, the curator of prehistoric anthropology at the Smithsonian, had this to say after examining the find with several other researchers:

> The authenticity of the skull is beyond any doubt. Its antiquity unquestionably great, to my own personal knowledge several such crania were discovered in the Hopewell group of mounds in Ohio. The jaws were prognathous (projecting beyond the face) and the facial index remarkably low.

Lompock Rancho, California, 1819, an enormous twelve-foot skeleton

---

A report that was thoroughly investigated in November 2008 found that an older lady saw a gigantic skeleton dug up by soldiers at Purisima, very near the Lompock Rancho, where the natives determined that the skeleton was that of a god, and it was reburied by direction from the padre of the local township authorities. This short report reemerged with a broader range of details in 1833, and now various authors and websites are consistent with the same exact story.

It goes just like this: Soldiers digging a pit for a powder magazine at Lompock Rancho, California, hacked their way through a layer of cemented gravel and found a twelve-foot sarcophagus. The skeleton of a giant man about eleven feet, eleven inches tall was found inside. The grave was surrounded by carved shells, huge stone axes, two spears, and thin sheets of porphyry, which is a purple mineral with quartz, covering the skeleton.

This huge man had double rows of upper and upper and lower teeth and had six digits on all four limbs. The soldiers consulted with a local Indian tribe who were displaced Alligewi Indians from the Ohio Valley area. The Indians suggested that the find was of very important religious significance. It is when the local authorities decided to rebury the giant. Several giants of this size were reported a couple of times by the *New York Times*, including an enormous giant who was fourteen feet tall near Albany, New York, in the mid-1800s. Two other giant skeletal remains and were both at least twelve feet tall were in Jeffersonville, Kentucky, which coincidentally is in the Ohio Valley and was reported extensively by the *New York Times* on May 22, 1871, and another was reported in Barnard, Missouri, by the *Providence Evening Press* on September 13, 1883. Furthermore, another giant, which was thirteen feet tall, was unearthed in Janesville, Wisconsin, by the *Public Ledger* on August 25, 1870. Another fourteen-foot giant skeleton that was completely intact was excavated at Etowah Mounds in April 5, 1886.

Salt Lake City, Utah, eight-foot, six-inch giant found, August 27, 1891

An irrigation worker was digging a ditch about eight feet near the Jordan River for a project near Salt Lake City, Utah, when he uncovered a gigantic skeleton that was amazingly standing upright. The find was measured eight feet and six inches in height, and the skull was measured eleven inches in diameter. The feet were almost twenty inches long. The skeleton had a copper chain around his neck and had three medallions covered with very mysterious Egyptian hieroglyphics. Other items found were a stone hammer that weighed fourteen pounds and some pottery, arrowhead, and other copper medals. Archaeologists believe this Nephilim creature belonged to a race of mound builders. This was front-page news in the *New York Sun* on August 27, 1891.

West Hickory, Pennsylvania, 1870, eighteen-foot giant uncovered

---

The headline of this chart topper reads "The Cardiff Giant Outdone: Alleged discovery of a Giant in the Oil Regions." The report came from the *Oil City Times* in 1870. The eighteen-foot gigantic creature of a human was dressed in his full armor and revealed some startling anatomic oddities and a skeleton that reached a staggering height.

"They exhumed an enormous helmet of iron, which was corroded with rust. Further digging brought to light a sword which was measured an unreal 9 feet in length." The report continues, "A well preserved skeleton of an enormous giant. The bones of the skeleton are remarkably white. The teeth are all in their places, and all of them are double, and of extraordinary size."

This discovery was buried more than twelve feet below the mound when it was unearthed. The sophistication with the discovery of these mounds had a sense of royalty to them, and some of the great wars were said to have taken place between these Titans or demigods with the native American Indians and is laced with historical content concerning almost all the tribal cultures. However, most people don't believe in giants but rather scoff at the idea, suggesting that it's all just folklore that got mixed up and was sensationalized by newspaper journalists. This is the same exact narrative from the Smithsonian Institute, which is behind the nefarious agenda of education that we were taught that Christopher Columbus discovered America and the fraud narrative of our historical Western education that came directly from the parasites and the lying agenda that was orchestrated by the Western conquerors of the "new world." The archaeology regiment direct from the elites who run the Smithsonian and the Darwinism lie was also included to ascertain that human beings came from monkeys, and of course, the missing link will never be found because it's all a ridiculous lie to suggest that we all illegitimately came from pine-scum and from fish and somehow a soul-less monkey, and yet the pieces of evolution is complete science fiction by a thirty-third Freemason named Charles Darwin.

The archaeologists do not profess to ever look into the existence of any giants because they are effectively told to "cover your eyes with your hands because you see nothing," to "cover your ears with your forefinger because you hear nothing," and "cover your mouth because you will not say anything!"

This narrative is the party line and comes directly from the official doctrine from the esteemed institute of atheism, and it flies against truth. Hundreds of newspapers all across the United States have full-blown stories of actual evidence when the railways were being built and the interstate highway systems along almost every single state in the union that the reality is that giant human beings who are the Nephilim and the Anakim who were in America probably one thousand years before any of the Asian Indians crossed the Bering Strait to come into the new world. The diamond in the rough that was hushed up, as well as any of these true accounts, would most definitely be the Trinity Temple that was found under the Grand Canyon, along the banks of the Colorado River. Several treasures of gold and ancient Egyptian artifacts worth well over $100 million today is stored in a secret basement in the Smithsonian Institution and will never ever be put on any display, for it is to be hidden treasure trove for the deception of the deceiving maggots of our world.

I disagree because of the overwhelming evidence within, and there are seven different avenues to explore and to be as certain that the suppression of truth was sought by the Illuminati-owned Smithsonian as a major cover-up to hide any existence of the Holy Bible from any confirmation of its validity to accept the truth and to gain knowledge of our destiny and importance in our life:

1. The Native American mythology
2. Overwhelming genetic data
3. Scientific reporting
4. Hundreds of early excavation records
5. Firsthand accounts
6. Discoveries featured in newspapers
7. The local towns' historical records where giants were found

Let us examine whether these are fables or legends, or is this based on real accounts and of what our true reality actually is? Why has it been hidden from us? Would it not benefit mankind, in general, to understand that the suppression of truth only benefits the few and negates our very conscious element of what our destiny is supposed to be like? How could a tall-tale and phony narrative benefit anybody? They lied because it surely was a major piece of the biggest fraudulent puzzle of the biggest deception in mankind's history. They did it because they were able to follow the agenda that has almost all the resources here in our world. Quite literally, the well-funded agenda is the global elites who actually print money and own 98 percent of all the money on our plane of terra firma called Earth. Herein lies the problem: how does one ascertain true intelligence and cognitive dissonance if all that we were taught was an intentional lie? The sorry portion of what is a fact is that 90 percent of the faculty and professors are believing it. The teachers have been hoodwinked into a false reality because they are just like the rest of us in that they were poisoned into this system of pseudo-education that is spoiling the goodness of our expandable brains and limits us to fully understand fact from mere fiction. The biggest tragedy of all is it makes it exceptionally more difficult to have a spiritual ascension to greater heights of our true origins and what God actually had in store for us to begin with. How can you decipher what is education if it's the incorrect and a frivolous lie? The very word "knowledge" means, I know, or a known fact of truth. Wisdom from this knowledge is readily available to all who seeks truth and seeks the knowledge of what our real manifest destiny is.

The evidence is factual that a former race of giants were actually the founders of the new world. The Native American Indians were not here before them, and then Columbus came a couple thousand years later.

There are legends of giants in many different cultures, such as the Titans of Greece, the Sardinian giants, and Goliath, almost assuredly is factual. In fact, the belt of Goliath was found inside an underground cave dwelling near Israel. The Bible has several references to the giants known as the Rephaim, Anakim, Zuzim, Sepherim, and, of course, the Nephilim. A scary thought is written in Scripture, in Numbers 13:32–33:

> The land, through which we have gone to search it, is a land that eateth up the inhabitants thereof; and all the people that we saw in it are men of great stature. And there we saw the giants, the sons of Anak, which come of the giants; and we were in our own sight as "grasshoppers" and so we were in their sight.

The overwhelming evidence is that these giants not only existed in all the world but also thrived in America and therefore set up shop. A Sphinx gold Egyptian structure was found in the bottom of the Grand Canyon, where caves from these monstrous human beings who were cannibalistic and had treasures that were literally worth millions of dollars back when it was discovered by M.K. Kincaid. Since then, has been completely hidden and suppressed from any historical record to speak of. However, there are newspaper accounts for very noteworthy and accurate tales of this reality that has been completely abolished by the Smithsonian. What roles do these institutions played in this matter teaches us about the role of bias in all studies of supposedly academic or scientific nature? What would be the reason for not keeping this information in the public eye? When did we lost our collective memory of these extraordinary inhabitants of America and the rest of the world? The other incredible dilution of factual events is that most of the anthropologists and archaeologists that work at the Smithsonian today has no credibility of knowledge that these creatures were here as residents of our true world. It is surprising that humanity is more aptly prone to believe in over sixty alien species that occupy the thinking of what's real and what's not based on facts, but they believe it anyway! Just look at Baalbek in Lebanon with the enormous 1,200-ton stones that no crane or any heavy-lifting machines could lift today. Just take a peek at the advanced mathematics and uniqueness of the engineering of the Great Pyramids or the Atlantis stories from the past. Plato wrote extensively about the most scholarly people in history of man, unless this is just a fable. We live in an age where we are hypnotized by our own ignorance, acting as if the atomic energy and digital electronics are the height of human achievement, giving plaudits of ultimate human ignorance gone awry. One might call it hubris; wiser minds would call it cultural myopia and adolescent grandstanding. The myth and antiquity are all too real, and there were other ages as great as ours, and whatever we have accomplished was built on the shoulders of giants.

Now is the time for academia that is somehow disconnected from the elite-controlled narrative, especially the AIAA, whom I consider to be heroes of truth movement for taking on the monstrosity of deceivers, mainly the Smithsonian liars, who are fitted just as NASA and the other liars of preposterous education of untruths that continues to poison our everyday lives. It is time to investigate the big elite companies such as the Smithsonian who were caught red-handed in the middle of the Atlantic Ocean dumping skeletons by a European tanker several years ago.

Samuel Clemens, a.k.a. Mark Twain, surely knew about the truth-killers denying God in our everyday lives because he spoke about it at length, especially through some of his most famous quotations. He remains one of my all-time favorites, and here are two very resonating quotes from him:

1. "History is strewn thick with evidence that a 'truth' is easy to kill, but a 'lie' well told, is immortal!"
2. "It's a lot easier to 'fool someone, than it is to convince them that they have been 'fooled.'"

After the Civil War, the Smithsonian began to adopt a policy of excluding any evidence of direct foreign influence in the United States of any history prior to the fake history of Christopher Columbus discovering America in 1492. It was all part of an expanded policy, the agenda of manifest destiny, the atheistic Darwin theory, and the missing links that will always be missing because we all live in the times of uncovering the deceptions and lies from all of the world's governments. The Mormons during that time were very close to their heads being decapitated of this fantastic deception because they were convinced that the "giants" and all the pearls and full-armored attire with gold and the technical cultural knowledge that you will find throughout all of ancient Mesopotamia were the lost tribes of Israel, and it was found throughout North America and suppressed completely today. The most powerful king of the deceivers from the Smithsonian undoubtedly is Maj. John Wesley Powell, and his tenure at the Smithsonian was from 1879 to 1902. Powell was a geologist and explorer who led countless expeditions and surveyed most of the American West. Powell was asked to write a report on the history of the ancient tribes and their probable origins, and it was going to be the official party line for the Smithsonian for the next hundred years. The title of Powell's initial report to the institution's secretary, which took place in 1879, was called "On Limitations to the Use of Some Anthropological Data." It is so revealing that it's "a slam dunk" about the narrative of deception to hide any existence of any giants that would be the tip of the iceberg to absolutely annihilate the nonsense called Darwinism. Just listen to his report:

> Investigations in the department are of great interest, and have attracted to the field a host of workers; but a general review of the mass of published matter exhibits the fact that the uses to which the material has been put have not always been wise.
>
> In the monuments of antiquity found throughout North America, in camp and village sites graves, mounds, ruins, and scattered works of art, the origin and development of art in savage and barbaric life may be satisfactorily studied. Incidentally, too, hints of customs may be discovered, but outside of this, the discoveries made have often been illegitimately used, especially for the purpose of connecting the tribes of North America with peoples of so-called races of antiquity in other portions of the world. A brief review of some conclusions that must be accepted in the present status of the science will exhibit the futility these attempts.
>
> There has been much unnecessary speculation in respect to the relation existing between the people to whose existence they attest, and the tribes of Indians

inhabiting the country during the historic period. It may be said that in the Pueblos discovered in the Southwestern portion of the United States and farther south through Mexico and perhaps Central America tribes are known having a culture quite as far advanced as any exhibited in the discovered ruins. In this respect then, there is no need to search for extra-limital origin through lost tribes for any art there exhibited. With regard to the mounds so widely scattered between the two oceans, it may also be said that mound-building tribes were known in the early history of discovery of this continent, and that vestiges of art discovered do not excel in any respect the arts of the Indian tribes known to history. There is, therefore, no reason for us to search for an extra-limital origin through lost tribes for the arts discovered in the mounds of North America.

Powell then goes on to definitely state that there are no foreign influences to be seen or studied in relation to the Pueblo- and mound-building cultures of the Americans that are believed to precede the Indians. In relation to his dismissive comments laced with lies and definitely an agenda that was ongoing and saying that there were no connections to the "lost tribes from the old world," it's also very interesting to know that Powell was the son of a preacher in Palmyra, New York, who had lost his flock to Mormon missionaries. The three main reasons about the wrongheadedness and deception of what he had suggested were

1. Powell was a champion of evolution. It's obvious that he was paid a boatload of money to lie;
2. the second theory would be called uniform gradual history, which makes little sense because it would mean that the Earth would go for huge spans with no catastrophes, which the opposite is actually true; and
3. the land-bridge theory, which states that the Indian tribes reached America from Asia across the Alaskan land bridge. Dr. Ales Hrdlicka was the first curator of physical anthropology of the Smithsonian but has been proven to be asinine, absurd, and an unprovable theory.

The (RH) negative blood Rhesus is a main characteristic belonging to the children of the fallen angels, the Nephilim.

This rare blood group only belongs to 14.8 percent of the world's population and is nonnegotiable not only in thought. The ruling elite marry into this group because anyone with (RH) Rhesus positive is not compatible with (RH) negative, and case in point are the wives that the Royal family chose are all of RH negative. Otherwise, the babies born more in likely will be deformed. The confusion about the origin of the word "Nephilim' and what it actually means comes from Hebrew, and it literally means "to fall," as in "fallen angel." The Bible, the Talmud, as well as the Koran (called Djinn) all talk extensively about the fallen angels. The book of Genesis relates that the Nephilim produced children (Genesis 6:4) with humans. Another fact that was hidden for hundreds of years was that one of these fallen angels killed Japheth, who was Noah's son, and fathered children with Japheth's wife after he was murdered. The true story was that eventually, the population was in parts of Northern Europe, especially London, and the Royal Family at Buckingham Palace. Many experts cannot make a conclusive narrative of exactly why the makeup of 85.2 percent of the rest of us are positive especially since the antigens in the negative "blue bloods" are quite unusual and sometimes has been characterized as nonhuman. One thing that these blue bloods have in common are the following:

1. They have higher IQs than the RH positive normally, but not necessarily, but normally are. They usually have more sensitive vision, and their other senses are usually more astute than RH positive.
2. Their body temperatures are lower, but their blood pressure is usually a little higher.
3. They definitely have better abilities of intuition.
4. They are better equipped to have psychic abilities.
5. They usually have red or brownish-red hair and have predominantly blue or green eyes and sometimes hazel.
6. They get sunburn more so, for their skin is more sensitive to heat and sunlight.

7. They cannot be cloned and have an extra vertebra, which appears to look like a serpent wrapping around the regular vertebrate. The Bible does tell the story of Noah's son being murdered by the Nephilim. The most common human blood-type is O and is a universal blood type. The RH factor is a protein found in the human blood that is directly linked to the Rhesus monkey. When blood type is inherited from your mother and father, it is known that this factor element of the blood is the most consistent human or animal characteristic passed on to the offspring. There are very few aberrations. It rarely changes. Most people have RH positive since the antigens are so very different, and places such as France and Spain are two of the more prominent areas that have a higher concentration of RH negative, especially with the Basque people, who are present in the area. The original group is obviously the Eastern/Oriental Jews, and almost 40 percent of Europeans are in this group, but it is a pretty safe bet that the source where this negative blood group originated was when the fallen angels who were expelled from God's kingdom came down on Mount Hermon. The land of Israel and Lebanon is where they perverted most of the seed of mankind. In the book of Enoch, it is exactly the reference that detailed the Nephilim and the fallen angels from God's grace and fell inter-dimensionally back to wreak havoc on the innocents of humanity.

The only time in human history that the spirit world propagated a wrath against humanity because of its revenge that was planned against God himself and took women for wives! In essence, the only time that the spirit world was able to have sex with the flesh world!
—Mark S Hollander

The New age Nephilim: They are the liars of the purported UFO-cosmology age of deceit and deception.

My personal transformation from being into a twilight zone of cloudiness in thought and of my true awareness of who I am, where I am, and who made me and are we in a unique reality to serve our own purposes, or are we in a precarious state of being in a different reality serving something or someone who contradicts the truth of who we are, and where we came from, and why are the liars of the world doing this to humanity? My whole revelation of the liars' secret antithesis of what our true world reality was is shortly before I took my wife and two of my three daughters on a seven-day cruise through the Alaskan Inside Passage. About a month before we flew from New Orleans to Vancouver to board a Disney World ship called the *Wonder*, I was immersed in a revelation that "Nephilim giants of renown" were uncovered throughout North America and scanned through true accounts that were reported (sometimes on front page) in hundreds of newspapers across the United States. Almost every discovery was turned over to the Smithsonian Institution and their archeological handlers, and furthering my interest was the discovery that every one of the accounts of giant skeletal remains were all of a sudden disappearing from any and all official records. The fact that the stories were sometimes actual headlines throughout the country automatically put my mental state in a mass confusion of how and why would America's most esteemed and oldest philanthropic and archeological studies from the past were either being discarded, hidden, or destroyed from the rest of the populace unless the act was intentional as a means to cover up historical and accurate facts from our past human history. I didn't know how to do any proper research because I was still caught in my new God called Scientism, with the pope being anything and everything NASA. I was even under a spell that I now call the Star Wars delusion, which gradually turned me away from my true manifest destiny of what true reality is, and any knowledge that I had and spent way too much time on were the over sixty alien races throughout the world and not enough time for any goodness and purity that I now get regularly each day from God and knowing that he came back in the flesh because of just how corruptible that man had become, and now it appears that the days of Noah are coming back to haunt and destroy mankind, steal money first and then your soul. It was starting to become a picture that resonated in my thoughts. Now uncovering the "wolves in sheep's clothing" has become a daily crusade to search for and live for the truth of my existence, to research and uncover the deceptions and the lies of our very own government.

HOODWINKED INTO THE AGE OF INSANITY!

The Firmament

I ask anyone who would be open-minded enough to ask yourself one question. Suppose that your belief system that you were taught your whole entire life was not only a fabrication with nefarious and evil intent, but was also indoctrinated into the most esteemed universities of thought and was taught erroneously by these controllers of our very school of thought. If you were taught mostly lies in school—and I do know that you have to understand that if you are not being taught what is *true* and what is *truth*—then what you have been taught especially through the world of physics of theoretical science based on assumptive conclusions and guesses based on a hypotheses has inadvertently has trumped over the world of an affirmative reality, of what is known! Somehow or another, dark matter and antimatter in our clouded world of only knowing a little less than 5 percent of what physics actually is all about, which leaves us with 95 percent of not even knowing what it is or what is known! This tells me that true experiments from the recent past scientists and the proven methods of providing credible experiments to ascertain truth or fact have invariably been replaced by psychedelic unprovable science based on theory. Have we all gone wild? I believe the parasitic intent was to distance your consciousness away from your creator and to prevent knowledge from being obtained. If you're not being taught what is true, then obviously, the attempt is to create a fantasy of cosmology that has created a "Star Wars illusion" that effectively has made humanity a pasture of nitwits and sleeping sheep.

I believe the "great awakening" has arrived, and the Freemasons hold the cards of deception that has invoked the biggest threat to their continued control is called flat Earth! Please do not take me at my word. I am only encouraging anybody to please just research flat Earth and come out as you will. There certainly is another side, and as long as one does proper research, then you will at least know why the so-called dumb-ass flat-Earth people aren't so dumb after all! It is imperative to do your own research! I believe the thirty-third Freemasons, the world's oldest ancient secret society, are hiding the very shape of our world where we all live, and I certainly believe that we are living the reality that they have placed us in. It's a false-reality especially because it's based on lies and more lies! They have spiritually killed our creator through science. Science is the world's biggest religion and is always 100 percent the opposite of what our creator says in the Holy Bible. Who do you believe? What God said in the Bible, or what man says? Man are known liars.

The thirty-third Freemason Nicolaus Copernicus was the hero ultimately for the beginning of the greatest deception in the history of man, the Copernicus heliocentric model, that he didn't want to actually release his works of pushing the agenda of the IHS/Jewish Banking families to suggest that the Earth is spinning 1,100 miles an hour, or 19 miles per second, and going through the Milky Way galaxy at 66,600 miles per hour and is exactly the opposite of what our creator said in the Bible. Copernicus did not go against the establishment because he knew that it is possibly artificial and not true by nature, only in concept, and was therefore a theory. Copernicus didn't want to release his works, and his works weren't released until after his death in 1543. He knew and was afraid and threatened to be charged rightfully so as a heretic, and he did dive into black magic aspect of witchcraft. His impact on pushing this agenda was monumental, and the official church position eventually weaned over time and the insane nature of purposely putting Earth into a demoted position from being the "center of the universe," as God said in Bible many times. We were unique in the cosmos for the first 2,500 years, when the Egyptian, Hindu, Norse, Mayan, Incas, Hebrew, and the Navajo all believed and knew that the world was flat and the dome over the world was the roof that God said he built in Genesis 1:7. It seems rational to me that these ancients knew something more than we do today, and even the old Hebrew conception of the universe knew. The maps and drawings were definitely unmistakable. They knew the Earth was flat, and it didn't change until theories were accepted into the mainstream of the cattle, meaning us, the people. Tesla himself was adamant in his declaration of scientists relying on mathematical equations instead of real scientific experiments. This new thought wave of propaganda is called lies, and by no means can humankind evolve into the very best that we can be if we were taught to be senseless sheep. Accepting these lies into our belief systems has created a huge vacuum of intelligent thought and closes the avenues for your spiritual thought to connect with your own sense of reality and to reach the vast potential of creatively expanding your wisdom and knowledge that ultimate truth provides to humanity.

NASA brought the grand deception into the twenty-first century and has created a dulled human state of mind and rendered most of humanity into a "Star Wars allusion" and to believe in space that other than lower Earth orbit, has no substantiation to even exist. I have had it spending trillions of dollars on a space program that provides us with their newest CGI pictures, and faking space is what these wizards of deceivers really are. The riches of their money not only puts you in la-la land, believing we came from monkeys and live on spinning ball rotating at 1,100 miles at the equator, which is absolutely insane science. One picture, one real picture provides all of us with something no one has ever seen!

We live in a Newtonian world of Einstein physics ruled by Frankenstein logic! I want everyone to understand that the book of Enoch was taken out of the Bible to hide the actual agenda, and if I may, these are the skeletons in the closet of our rulers, which absolutely is the global elites. Their number is around four thousand.

PART 1: WAR IN THE THIRD HEAVEN

Enoch, the son of Jared, was taken up to God without dying (Genesis 5:18) He was the father of Methuselah. Enoch lived in such close fellowship with God that he was translated into the presence of God without dying. Hebrews 11:5 attributes the translation to faith. He was the great-grandfather of Noah and the twelfth from Adam. The extraordinary story of Enoch and the very reason that the Jesuits-Vatican ancient Jewish Tribes dislodged Enoch from the Bible was because Enoch's ascension into the heavenly realm and the information that he revealed about the war in the third Heaven was because the book of Enoch uncovers the skeletons and truth of the rulers and ancient descendants of the Rothschild's, the Rockefeller's, the four thousand people who actually rule the world as we all know it today.

Two hundred inter-dimensional angelic beings descended from the major revolt in the heavenly realm and were cast down and upon Mount Hermon. They came down to a village that numbered over a several hundred, and the sons of God are what they are called, and the perversion of the purity of mankind will forever be dramatically changed.

Once upon a time, there was a "godhead trio" consisting of God, the word, and God the Spirit. They existed before time, during a time, and then after time, which certainly sounds paradoxical, but our level of understanding and comprehension can only think within our dimension. It is the only time in our origins of history that the supernatural inter-dimensional beings mingled with the flesh of mankind. The encounter was documented extensively in the book of Enoch. However, the nefarious agenda of these beings was to vomit on the purity that was man, and they took wives and produced supernatural children called the Nephilum. In Genesis 6:4, it was taken out by deliberate means, and it is astonishing that Enoch lived for 365 years and, according to the Old Testament, was the only human being who never died! God said that he wanted him and took him to paradise.

These beings came from the kingdom near the throne where the creator is stationed called Heaven, or the kingdom of the moist high. They exist outside of our dimension. I can chronicle it more in a scientific analogy that also charts the difference between physics and quantum physics. Physics is the flesh realm of our everyday reality, whereas quantum physics is almost essentially in the spiritual realm. The fact that the world's best scientists only know about 4.7 percent of what physics actually is as far as any scientifically verifiable proven experiments to ascertain reliability to provide an analysis of factual truth suggest to me personally that physics is mathematical convoluted and theoretical lies and propaganda. The other side of this spectacle of delusional science is what we call the spiritual realm, which is quantum physics.

Before mankind was created by the word of God, he created the angelic realm, which were beyond numbers. The one-third who rebelled led by Lucifer were kicked out of the third Heaven. The major premise of Lucifer's ambition was to rise above the throne of the Most High, being the father's. Once mankind was created, he first deceived Eve along with Adam in the garden and thus became the ruler of this world.

The Word of God placed a curse on him and punished Adam and Eve for their disobedience. Lucifer tried to circumvent God's curse and set out to corrupt God's bloodline. Once Caine murdered Abel, God's bloodline continued with Adam's younger son, Seth.

This is where Lucifer dispatched the two hundred of his fallen angels to a place on Earth called Mount Hermon. Lucifer's goal was to prevent any of God's prophecy from coming to pass, namely, to have a savior for mankind. I will put enmity between thee and the woman, and between thy seed and her seed; it shall bruise thy head, and thou shalt bruise his heel. The Lord Jesus Christ bruised the head of the serpent when he died on the cross. He said, "Now is the prince of this world judged. Now is the prince of this world cast out."

The ancient Cabala-Hebrew for Qubballah is defined very clearly, and it seems the very narcissistic agenda is at play even in description.

1. A Secret Group of Plotters or Political Conspirators.
2. Club Group.

The Vatican installed Peter the Apostle as the first pope right around AD 300 and gave great empathy as pagan worshipping the saints who wrote the Bible, placing the saints intentionally to defrock the

---

A Catholic Priest recently died in 2016, and he claimed to have performed over three thousand exorcisms in his life. He once claimed "the greatest lie that the devil wants mankind to believe is that he doesn't exist."

The fall of Lucifer finds its origins with a fallen angel in Origen based on the interpretation of Isaiah 14:1 through 17:1), which describes a king of Babylon as the fallen "morning star," and the description was interpreted of that of an angel because the literal application to a human king has been pretty self-evident. The very image of the fallen morning star or angel was thereby identified as Satan in both Jewish pseudepigrapha and by early

Christian writers, which came after the transfer of Lucifer to Satan in the pre-Christian century.

Origen and other Christian writers linked the fallen morning star of Isaiah 14:12 to Jesus's statement in Luke 10:18, "I saw Satan fall like lightning from heaven" and finally to the mention of a cataclysmic fall of Satan in Revelation 12:8–9. In Latin-speaking Christianity, the Latin word *lucifer*, employed in the late fourth-century AD, gave rise to the name "Lucifer" for the person believed to be referred to in the text. Orthodox Judaism does not believe the name Lucifer is a reference to Satan but rather just a taunt against the King of Babylon. Christian traditions have applied to Satan not only the image of the morning star in Isaiah 14:12 but also the denouncing in Ezekiel 28:11–19 of the king of Tyre, who is spoken of as having been a "cherub." Rabbinic literature saw these passages as in some ways parallel, even if it perhaps did not associate them with Satan, and the episode of the fall of Satan appears not only in writings of the early Christian Fathers and in apocryphal and pseudepigraphic works but also in rabbinic sources.

It's so very unfortunate that the modern evangelical commentary on Isaiah or Ezekiel sees them as providing any information at all about the fall of Satan, and whether a fallen angel is wicked or rebellious isn't the relevance behind from which they came. The omitted books from the Bible on the origins of man was replaced by an agenda that I likened to turning a man or a child around in circles for over 474 years. The heliocentric global theory model from the works of Copernicus is the exact opposite of what actually is taught in Scripture, especially the old Hebrew and Greek text from the original transcript, which created a gradual vacuum of faith and understanding from what our creator explicitly said and how it is now isn't defined accurately as opposed to the way it was. The popular reason given why several books were passed on during the more modern version of the New King James text was that the book of Enoch was too frightening to be included in its modern-day teachings. My only problem with this is the fact that the other books taken out of the Bible such as the book of Adam and Eve, the book of Jared, were all part of perverted man's spirit from what the truth actually is defining. It is very hard to swallow a Merriam-Webster Dictionary from 2012 that doesn't have a definition of the word "firmament." I am convinced from reading the book of Enoch and the insurmountable attention that this amazingly bizarre story of the twelfth man after Adam and the great-grandfather of Noah was excluded from the

Bible because when the five major religions were established by the Club of Rome and the people at the top of this agenda were going to set religion up in order for all five religious orders to always be at odds with one another. In fact, to hate one another and create confusion and dissension and hateful to humanity over the promise of money or gold or a better life to escape being enslaved or brutally oppressed.

The fact that Enoch is the only human being who ever walked in this world who never died. God took him after his 365th birthday, but the very book of Enoch is all about the skeletons in their closet. The global elites who own almost 98 percent of the world's money made doggone sure that several books would be deleted from the Bible, especially anything that would be the transparency of where the origin of man came from as well as the corruption of man caused by the very serpent in the garden of Eden who tempted Eve to eat the fruit tree of knowledge and become a God. Well, this explains the revolt in the third heavenly realm, and it has led to all suppression from what the truth is. We are now living in the time of the final battle between man's adversary, but in a lot of gray area that has presented almost a case of man not being able to break free from the great deception that has so much plagued man's spirit conscious to reconnect with God and to reconnect to each other as being stewards to the truth.

Subject: THE FIRMAMENT

The word and the definition of "firmament" is as follows: great vault or expanse of sky that separates the upper and lower waters. God created the firmament on the second day to separate the "waters from the waters" (Genesis 1:6–7). One use of "heaven" in the Bible is to refer to the ceiling or canopy of the Earth. Heaven, in this sense, is also referred to as the firmament or sky. Into this expanse, God set the sun, moon, and stars (Genesis 1:14–18). In Genesis 1:6, the firmament separates the mass of waters and divides them into layers. The firmament is mentioned seventeen times (KJV), including Genesis, Psalms, Ezekiel, and Daniel. It is described as bright, transparent-like crystal revealing the handiwork of God and signifying his seat of power (Psalms 19:1 is on the tombstone of Wernher von Braun, father of NASA space program). Before we get into the massive deception from all of the world's space programs, we will entertain what I believe triggered this deception that has matched nothing else. This is the biggest lie in mankind's history and was planned for a very long time. In order to create cosmology, you would have to lift off the dome over our heads that God, our creator built. The study of alchemist Copernicus, especially to the teachings of Pythagoras from the sixth-century BC, was propelled more than one thousand years later to create a brand-new narrative of guessing the very hypotheses himself to create a spinning ball going around the Sun, which was placed at the center of an infinite universe. The heliocentric model could only be possible if the dome firmament is rejected and for it to disappear not only of thought but of actually purging it from Scripture, and that was done by the IHS/Jesuits-Freemasonry Cabal who would rather keep common man in complete confusion. That started after Admiral Byrd's last trip to Antarctica that was called Operation Deep Freeze, which went on from 1955 until 1956, and then all hell broke loose. It was a very secretive and paranoid time for all Freemasons who are more aligned with their secret society with their secretive agenda. Then they are with their respective countries that they lead. They are trying to hide something of such a mind-blowing magnitude of—may I say this—of Biblical proportions that, in essence, Antarctica became off-limits to all man and is the only country in the world that is owned by no one and only visited by the World Banking families and invited guests, mostly heads of states. Pres. Barack Obama visited two years ago, and astronaut Buzz Aldrin visited around six months ago and twittered out a scare for all the delusional Star Wars alien believers of the imaginary cosmos. He tweeted "about an evil existence that exists," alluding of an upcoming alien invasion rather a fake alien invasion. The fact that he got sick going to Antarctica and spent time in an infirmary and thought to have pneumonia made his trip not as poignant as he would have wanted.

I firmly believe that Admiral Byrd found the "edge," and I believe they did traverse in the most wintry conditions past the coast of the ice-wall barrier and found the firmament between 800 to 1,200 miles inward. Shortly after Byrd's trip, several very unusual events started to unfold, and it was very chaotic from the Cold War between Russia and the United States, but I am convinced that Admiral Byrd found the dome. I think literally the elite Freemasons went bonkers. One of the dramatic events was that the two superpowers started exploding high mega-ton nuclear bombing at the sky. This was four years after the birth of NASA and was called Operation Fishbowl. Other tests were conducted as well. Operation

Bluegill and Operation Starfish were interconnected with Fishbowl, which continued the high altitude Nuclear explosion tests. The attempt to get through that crystalline transparent glass ceiling proved unsuccessful, I must add. Operation Dominick followed, which eventually led to a moratorium on nuclear testing.

Some scholars argue that the Hebrews had a primitive cosmology where the firmament was visualized as a rigid, solid dome—a celestial dam (Genesis 7-11; 2 Samuel 22:8, Job 26:8 and 37:18, Proverbs 8:28, Malachi 3:10). Above the firmament flowed the heavenly waters. The firmament was punctuated by grilles or sluices, "windows of heaven" through which rain was released. Others argue that such interpretations are unsound in that they confuse poetic and figurative language with literal prose. Others say Israel's inspired writers used language of experience and appearance rather than language of precise scientific description. I have researched a lot of books and videos to apply knowledge to a pretty amazing photo during Shackelton's second-to-the-last polar exploration of going past the perceived narrative of going past the South Pole. I do not want to investigate anything that I know has no evidence of existence. The dome-firmament found by Admiral Byrd in 1956 in the Operation Deep Freeze that would turn out to be the last Antarctica trip for Byrd and was immediately closed off completely to the public. That would include no exploration, no oil companies, but was going to be headed by all the powerful governments that are outfitted with navies capable to patrol Antarctica twenty-four hours a day. Just to look at your common everyday wristwatch. Your 1 to 12 rounded and up-crusted edges is in fact a replica of Antarctica and is the true map of what our world looks like.

Auguste Piccard told Popular Science in August 1931 issue upon his ascent with his weather airtight into the stratosphere: "It seemed a flat disk with upturned edges." I did come across an amazing picture of Antarctica that was taken in 1922 by George Rayner, a photographer, that was found online in the Museum of Victoria and is an actual picture of the "dome firmament edge." Just as God the creator said, he will separate the waters from above with the waters below. The picture is so very freaky cool and is the "epic picture that stands alone." An earmark comment from Hollander himself, concerning Rayner's photo:

> "The Rayner photo of Antarctica is a clear unobstructed from any technological submission of modern C.G.I. to

doctor or plant an image into the photograph just as the "blue marble" was and has been debunked void of any air of reality. It is officially a 1922 and its "authenticity" that was impossible to defraud through the implementation of techniques like virtual reality and the list is multiplying to fake our new reality. The 666 machine of our new artificial existence called the super-computers has convinced most of us of the results to keep us in a senseless state. Because we all are living in an pseudoscientific age of allowing psychics, in which we know very, very little about, a little less than 3% of what our knowledge is on this but we do know what electromagnetism is and our universe or in that we are made of water and electricity, just as you see in lightning and the waters of the oceans and all that we experience. but this nihilistic Agenda was indoctrinated into all a parasitic reality that is not our own. -

in its "authenticity" of our very own reality because in 1922, you couldn't doctor and fake pictures because the 666 machine called the computer helped speed along a fake reality that has become a conscious of realism that was created from artificiality.

The Van Allan Radiation Belts are nothing more than the Dome Built by God. (Brett Salisbury, ex-intel CSN G3)

There are two different sources of light that our viewing eyes see each day when you see that neon and all the noble gasses glow when they are being bombarded with electrons. Therefore, I conclude that daylight and sunlight are completely separate from each other. The sunlight is yellowish and white, which is called direct sunlight and is the cause of shadows as well; but the second light of light that you receive is called diffused light, which refracts in the lower atmosphere, formulating many visible colors. The atmosphere is consistent of at least five layers.

1. The Troposphere: The lowest point of the atmosphere. It reaches approximately 29,000 feet and is essentially from the ground of Earth and upward to the clouds. The jet streams high-pressure

systems circulating or merging with low-pressure systems, and any meteorologist studying the weather patterns and unfortunately, we should have several cameras such as the Hubble Telescope that would be more than capable of taking a 24/7 monitoring Earth in its total globe form spinning at the axis at 23.4 degrees and have not been able to provide us with a 24/7 sustainable footage of Earth and just imagine the tremendous benefits that Meteorology technology. Instead, we get computer-generated images that have been predominantly used to confuse mankind, and not a single image is real. How can anyone believe CGI and paintings of Earth and the imaginary spherical planet system from a dogmatic system that started five hundred years ago. It was essentially unchallenged, but the poisonous doctrine of theoretical physics were taught to us by mostly liberal professors who were well-versed in alchemistic theory developed by math equations to try to prove the unprovable element that physics brings in. It was from the alchemy teachings of the occult who forwarded the agenda with this form of witchcraft and black magic that this teaching was thwarted upon us as children and innocent adolescents, but none of us knew better. We just didn't!

2. Stratosphere: This is the area that is thirty thousand to forty-two thousand feet up and is certainly the second atmospheric layer where all airliners fly because it's the most stable layer and is less affected by the electromagnetic particles that causes a lot of dysfunctional wind cycles of the reflected molecules. It comes from the Sun's rays and gets magnified from the dome-plasma roof in the sky, the rooftop of our world that God himself created. The only jet craft that would ascend over forty-two thousand feet normally are of the military variety. The second layer is also where the ozone layer is, which absorbs the ultraviolet rays that emit very high radiation and cause skin cancer. It is the area where you are when looking out of the window of a Delta Airlines flight.

3. Mesophere: It is almost like the middle layer in this part of the atmosphere, about forty-two thousand to fifty thousand feet. It's as if it's a conduit between the heated particles of the uppermost atmosphere and the lower stratosphere and troposphere that reaches the ground that we stand on. It is a conduit that cools off the heated falling stars or meteoroids that modern science is

determined to provide a different narrative to what these falling stars are coming from deep space. In my humble opinion, it is an outright lie that keeps the masses in line with the continued deception of Earth being a speck in an infinite universe.
4. Thermosphere: The part of the upper atmosphere that forms almost like a donut of figure eights across the plane of our Earth.

This is the area in the upper atmosphere that forms what science calls the James Van Allen Radiation Belt and makes any space travel essentially impossible. The highly invisible field of highly charged electrons with temperatures that will get from 1,200 degrees Fahrenheit at 110 miles to 1,200 miles serves as a barrier of the dome firmament where temperatures at 158 miles up can get to an astounding 4,675 degrees that will omit any kind of space travel in our reality of common sense, and it doesn't seem like common sense is conclusive in the narration of the Freemasons' space programs of the world. The protons and the elements receive deadly ultraviolet radiation from photonic light and heat from the Sun. The Sun never does rise or fall past the Tropic of Cancer to the Tropic of Capricorn for the June solstice and the December solstice. If one uses the same calculations though a whole lot closer, we can clearly see the Sun and the Moon is only about 70 miles wide, as evidenced from the solar eclipse in August 2017. The Moon shadow was only 70 miles wide from Charlotte, North Carolina, to Nashville, Tennessee, and we were taught that the Moon is a spherical body and is 2,159 miles wide in circumference. The shadow only being 70 miles wide almost verifies that the Sun and the Moon are much closer than what they are telling us. Much, much, closer, probably based on the calculations that Leonardo Da Vinci himself calculated to be about 4,000 miles away. I often see the Sun going down in the west part of the lake Pontchartrain causeway bridge. At 24 miles long, it is the longest bridge over water in the United States. While the Sun is descending away from us out of variance of view due to perspective, you can see how it actually shrinks into no bigger than a pencil eraser. As it peers down on the horizon and makes a localized sunspot over the lake, the crepuscular rays highly suggest that the Sun is also about 4,000 miles away!

The Thermosphere can also be split into the Ionosphere and the Exosphere. The plasma roof dome acts just as a "mirror" and creates the colors from the rainbow. This altitude is the mountain top of the plasma dome, which is 7,200 miles up, and the aurora's, with the glowing lights, are usually found in the polar regions such as Norway and Alaska. When I took my family on the Alaskan cruise in July 2014, we didn't see the "Northern Lights," and it's because the furthest North that we sailed to was Skagway. We would have had to go to the mainland of Anchorage, which isn't the front-row seat of this beautiful, wondrous experience. You would have to venture 358 northeast to Fairbanks to see the continuous beautiful aurora called the Northern Lights. Norway is another polar region that experiences the refracted light from the roof of the night called the firmament. The reason that you get nature's show at the polar regions is because of the magnetic field that gets stronger than at any point in the world. The north magnetic pole has a fixed star called Polaris; it hasn't moved in about 3,500 years as viewed by early man. The stars revolve around a stationary fixed Earth just as the Sun and the Moon would, but the north magnetic pole star does not move, and the power of electromagnetism sensationalizes and excites with interactions with gas molecules and forms into colors called the aurora borealis, one of the most beautiful sights of the beautiful plane called Earth or terra firma. Another anomaly that is caused by the roof of the night are sprites/elves, which are created from Earth's highly charged glass sky, and we clearly see these Sprites in Antarctica at night mostly during the summer solstice. During the winter solstice, when the Sun is over the Tropic of Capricorn, and Antarctica is experiencing the twenty-four-hour Sun, which smashes the globe as bad as Jeranism shows in his intro to the YouTube channel called Globebusters.

On June 28, 2015, from a launch of the SpaceX Falcon 9 rocket launch, it looked pretty convincing that the rocket crashed into the plasma dome. Everything that we see such as the auroras, halos, sprites are all occurring at the glass ceiling. All the FakeX and NASA rockets will definitely go below the Karman Line, and all the rocket technology is simply unproven to be factual. In fact, when they allege that man is in these rockets launching into low-Earth orbit has been proven to be a falsehood and so very easily proven to anyone who does proper research on this. In fact, six of the seven astro-nots from the Challenger explosion in 1986 are still alive today. One is a professor at the University of Wisconsin, the commander Michael J. Smith, and also holing down a professorship at Princeton University is the school teacher named Christina S. McAuliffe, who now uses her middle name Sharon McAuliffe. These rockets will never go straight up to space, and every rocket launched at the Kennedy

Astronaut planting waving flag on moon.

Space Center will take an easterly turn at approximately fifty miles up and eventually ride the inner space until it crashes into the Indian Ocean and clearly out of view from anyone because after the Apollo, Space Shuttle launches away from your visual perspective. The nervous people at Johnson Space Center Mission controls in Houston, Texas, in just a matter of ten to twelve minutes, will get a convoluted and exhilarating message that "the astronauts" made it to space! We have to take their word for it even at the command center in Houston. It's amazing just how long people have been fooled and are in no way can anyone ever see anything real going into space. They feed you simulations of flight. They also send us animations of how the flight works and the rocket boosters being jettisoned back to Earth, but we have never seen any manned rocket go to an alleged space, which seems literally impossible. All communications from NASA astronauts are just land-based, and so are the satellites. Satellites are simply not real, and the fabrication can be proven just by the impossibility of satellites flying in the Thermosphere where temperatures would melt any kind of flight based on the composition of what rockets and satellites comprise of. They can send a drone up, and of course, this has been militarized as a weapon of war. These rockets normally pancake out at three hundred kilometers.

I almost named my quest for truth in this manuscript "the firmament," the dome that, in Genesis 1:6–7 and 8, God said, "Let there be a 'firmament' in the midst of the waters, and let it divide the waters from the waters." Thus, God made the firmament and divided the waters that were under the firmament from the waters that were above the firmament; and it was so. God called the firmament Heaven. I was so very amazed at how powerful my personal spirit seemed to literally take over my life and the physical pain that my car accident created, and the financial strain that ensued and came about from missing more than a year of work created a brand-new element in my personal life that to this very day has been astoundingly amazing. The severe pain that I felt and the countless sleepless nights because of the vibrations that enveloped into my legs in this case of full-blown myelopathy made comfort an afterthought. I cried to God each night to please help me alleviate my pain and allow me to sleep. It was transcending my spiritual consciousness in an accidental way. I was having a myriad of problems and confronting them to my wife and to my boss at work and convoluted any thought of any rationality to schedule the spinal fusion surgery and to also have part of my prostate that needed to be taken out as well. I wanted to write about the classical dome roof that creates a matrix of being in an enclosed system that all cultures from the past seemed to ascertain that very thinking of our reality.

The University of Colorado, from a report in Science Daily news (sciencedaily.com) that occurred in November 2014, found an invisible shield that was approximately 7,200 miles above the Earth's surface. In essence, if you just look and research the old Hebrew conception of the universe, it just verifies exactly what is written in the Bible. An invisible shield of plasma material with "killer electrons," which whip around the (plane) of Earth at near light speed and have been known to destroy and degrade satellites or empty rockets—sorry astro-nots—and it is virtually not just improbable but also impossible for any space travel to even approach this plasma shield that the world's best scientist accidently found. I even find it amazing how our very sense of knowledge of what the Bible says and proves that it is the final word for any ascertained thought to form a favorable income of what is verifiable thus is consistently true. It is goofy to realize just how idiotic our knowledge is, being stuck in a matrix of lies to invariably not know what your true destination is because without the absolute truth of one's reality actually is, there is no way to unpurge these lies of educational programming. The "Star-Wars delusion" coincides with Darwinism. Considering why the "missing link" has never been found, it's because it does not exist. Evolution is such a pitiful lie, and I often wonder how to explain the human soul, the epicenter of these neo-maniacs, to assert the lies of your true destiny and steal your soul from the center of their intentions. The whole agenda is front and center for the world to see, yet few are seeing it, and it is sometimes fuzzy for all of us, so unfortunately, the process remains kind of slow, but the breeze of deception and its suppression of allowing man to expand his own horizons with an expanding mind and foreboding wisdom to reach his heights in his or her life through an expansion of your mind, to free your mind from the lies and some of the belief system that is hard to subtract from your thought process or belief system. Please do your own research! Remember what a wise doctor in the person of Dr. Wayne Dyer, once suggested, "The ultimate ignorance is the rejection of something you know nothing about yet refuse to investigate."

It's very unfortunate with the newer generations coming up. The Jesuits are always hard at work to purging the Bible and have taken out very important words to absolutely dull your connected sense of spirituality to even know what the heck you are even reading. The word "firmament" has been deleted to include the word "expansion," or "expanse," which, in itself, confuses even me. That just as Bill Nye, the science guy who is always representative for the science doctrine for Disney and is a Freemason and the other 33rd Freemason, no other than Hollywood-NASA-Disney official spokesperson for the grand deception, is Neil deGrasse Tyson, and these two fairy tale heliocentric spinning balls spreading their web of spinning lies are usually together as "two itsy-bitsy spiders," especially when they're sprouting their nonsense on the globe-dominated shows. Tyson's official proof of the existence of gravity was laughable on all issues with no assertion of proof other than dropping a microphone to allegedly prove to his TV audience his version of the proof of existence of gravity. I didn't want anyone reading this as sacrilegious, but religion is part of the problem! I was raised a Roman Catholic in the suburb of Metairie, which is six to seven miles from New Orleans, and when I thought that I found a new reason for living when I inadvertently was saved by the blood of Jesus Christ in the midst of a contentious four-month marriage that involved my eldest, a custody battle erupted at the end of 1991. In my depression of not having a normal relationship with my eldest daughter, I found Christ. As I look back on it now, it was a blessing as well as a curse. The only time that man or woman finds a new beginning in Christ is when he or she is at the lowest ebb of uncertainty in life. The peaks and valleys of our lives will always be constant, and the learned lessons of the past will spring a better future, and the more meaningful soul represents your new nature of your worth, to your God, your family, your friends, and even your enemies. There is only one truth, and anyone who cannot question whether his belief system could possibly be wrong is just accepting things that were never ascertained as being factual or even beyond what a theory represents or not represent.

If the spirit of man dies, and he does not pray or does not recognize our creator when all is very good in his life, but when tragedy unfolds, he cries to God for help. Our spirit awareness and connectivity to God is only when we need him the most, when all is horribly bad, and we almost never ask for any blessings or continued blessings to for our lives. I don't wish to go full fire on any connotation that I'm a religious fanatic because *I am not*. In fact, I belong to no church's or any denomination of any kind. Am I spiritual? You're damn right that I am, but I didn't decide that I am pushing any faith of connectivity to any religious faith of any doctrine that any of these faith's serve. I fully understand the cloudiness of any understanding

Lunar landing approach.

what the word of God says now is totally different than the old Hebrew and Greek text of what the Bible said then. Man even went as far as corrupting the very word of God. It emphatically states in the book that it's a word that will stand against any and all times. It would be an abomination to pervert it just as man has done. The Vatican has utmost controlled the doctrine of the word because of its enviable place of Christianity of man's power to be corrupt that it came about that these parasitic controlled governments that created the worlds five major religions: Judaism, Hinduism, Christianity, Islam, and Buddhism. Mostly, these five religions were built doctrinal by man to hate the other four religions that were undoubtedly started by the collected forces of money and natural power to control the masses with a unified message from the puppets that are in backed by the ancient ancestors of the Jewish banking families with the IHS/Jesuits from the ancient days with their own army of bodyguards called the Knights Templar. The head of the Vatican, the Roman Catholic church, have over one billion constituents that are so blindly hoodwinked and do not know better. As a Catholic myself in my childhood, it was the time during the false conception of my own reality and the poisonous indoctrinated lies were placed in front of me, as it is placed in front of everybody who gets educated into this doctrine of deception. It's very convincing to me that this plan has been in place for a very long time, a lot longer than the 474-year lie about the heliocentric Luciferian spinning ball around the Sun. It's no accident that they are pagan sun-God worshipers, and the model was as exactly as planned, and other Freemasons had been pushing it. They also knew that it was going to take a very long time to execute the deception, and the building blocks of lies that thirty-third degree Freemason Sir Isaac Newton, with his theory of gravity, which even in June of 2017, cannot be explained, much less proven to even know what it is? When Neil deGrasse Tyson—actor, physicist who has been pictured with his 33rd Freemasonry Illuminati images of pagan sun-God worship and is very well paid to continue the theory of curvature and to stomp out any flat Earth truth, yet on the Joe Rogan show—the greatest TV poison bullshit scientist in the whole world couldn't explain what "Gravity" is. Amazing, isn't it? He furthermore in his interview with Joe "would sell out his own mother" to see his delusional space craft from out of his cosmology imaginary space! Who in their right mind would even suggest to sell out your own mother? He told the whole world that seeing an extraterrestrial UFO is essentially more important than his own mom. I say when you die, Mr. Rogan, make sure you take your money with you because it probably wouldn't matter, as it's exceptionally warm and hot where your soul is going! The agenda itself has successfully suffocated the truth so much from the whole world that the very acts of war and the ravenous cost of lives and countries that are burned to the ground are then rebuilt with an olive branch of peaceful intentions from the very people who finance all sides to all wars since the 1700s. Napoleon's French army and its very resistors were all financed by these neophytes.

The birth of the final deception of man: the Copernicus heliocentric model of cosmology fairy tale

----

Nicolaus Copernicus, a Catholic astronomer, mathematician scholar, scientist, and a religious figure, was born on February 19,1473, to Nicolaus Copernicus Sr. and Barbara Watzenrode. His father was a very affluent copper merchant from a very wealthy copper businessmen in Torun, West Prussia. They say he is a Polish astronomer, but real facts say that Copernicus was German. In fact, German was his first language. He is one of the main missing pieces to create reality from an illusion or a theory that going in they knew didn't actually exist In 1491. Copernicus entered the University of Krakow, where he studied painting and Mathematics, but his passion was the "cosmos," and by the mid-decade, he obtained a doctorate in canon law. It was during this time when he perfected alchemy and was in Freemasonry during this point of his development into a pawn who was funded remarkably well by the owners of the whole chessboard that we call Earth! I often wonder why would he not want to publish something that he truly believed in, but despite the parade of the threat of jail, he was very affluent as a result of his works even though in public he took a conforming role of obedience. In private, he was an accomplished alchemist, which wasn't really popular then. What we know about it can't be accepted today either. The black magic and the witchcraft involved is a real thing that people seem hard to digest and was instrumental in establishing the final deception on the agenda of the Luciferian global elite's plans on stealing the souls away from God. Some astronomical hypotheses at that time, specifically epicycles, and eccentrics were seen as mathematical devices to adjust calculations of where the heavenly bodies would appear, not an explanation of the cause to these motions, only proved through the use of equation methods.

The Copernican theory took six decades after his death before it started its ascent to a different reality based on a wild theory, in essence, placing the Sun as the center of the universe rather than the Earth. The publication of his book *De Revolutionibus Orbium Coelestium* (*On the Revolutions of the Celestial Spheres*) just before his death. *From the Copernicus Complex: Our Cosmic Significance in a Universe of Planets and Probabilities* by Caleb Scharf, and it, I can see a lot of conclusions that apparently sneaked in as being a fact, even though when all so-called methods of determining an official hypotheses, we still get guesses for answers instead of ascertaining true fact. None of the findings that I have been uncovering in his work do I find any representation of truth being proven. Let us chronicle a few snippets of just exactly what I am referring to. A hypothesis for the early stages of our solar system's history was put forward by the dynamicist David Nesvorny, and it argues for our system to be more active, less unusual, and less significant. In this picture, the young solar system contained five, not four, giant planets. In my opinion, this was suggested to make the formulated equation work to fit the model. The suggestion that a fifth planet could have been an icy giant, perhaps intermediate between Neptune and Uranus in mass and in an orbit somewhere between that of Saturn. The formation of an object like this out of a gunk of gas and dust around our baby Sun is certainly plausible and could add more spice to our solar system's orbital history. Nesvorny's simulations of the subsequent evolution of such a system typically result in the fifth giant being given the gravitational heave-ho by Jupiter, which ejects it all the way out to interstellar space. The arrangement of our major planets that's left behind in such simulations is often a statistical match to the configuration that we see now. In other words (and perhaps counterintuitively), the presence of this extra planet could be just what the doctor ordered. Having had a fifth giant planet now lost seems to increase the likelihood of our youthful solar system ending up looking like the present solar system. This is certainly interesting twist and a great reminder that we still don't know exactly what happened four billion years ago in our own system. Perhaps the current, rather-subdued dynamical state of our planets owes something to a rather more violent, hotter, dynamical past. Perhaps we flung a sister world out into the void. The brutal indifference of natural selection may work on planets too. It seems like a good bet that our solar system belongs to the 25 percent or so of planetary systems that have never been particularly chaotic in the past. We now think that planets in the middle range, from super Earths to baby Neptunes, are among the most numerous types of planets of all, perhaps outnumbering giant worlds by a factor of four or more. I did read the whole book, and my golly, the study of astrophysics are all guesses, nothing but dribble on whatever floats into your imagination. Make sure you can be accommodated by subjection of all your wordings, such as, perhaps, and we love to try to prove our ideas!

Okay, Nesvorny was suggesting that the age of the big bang was four billion years ago, and then the Fermi paradox suggests that the Milky Way galaxy is at least ten billion years old. He also thought that interstellar travel would be slow, taking thousands of years to go from one star to another, and the thinking was ancient species have spread out everywhere. The ridiculousness of this unproven science reminds me of a specific text that I have sent out to thirty-seven of my supporters that continuously support my efforts and often indulge me with questions concerning my work and research.

> Family and friends, think about this! If you boarded a flight on a 747 airliner and went to Mars at five hundred miles per hour, it would take you thirty-two years just to get to one of the wandering stars called Mars, if it's really that far away, which is BS. Think about it: sixty-four years there and back (Frankenstein space logic).

To create other illusions, the famous American scientist Frank Drake, introduced the Drake equation in 1961; it focused on the discussion and search for life in the universe. The factors include the fraction of planets capable of supporting life and the length of time that civilizations might broadcasts their presence. The discontinuation of the Big Ear telescope that actively participated in the SETI development to try to hear for any radio signals coming in from the cosmos was discontinued in 1998 and again is the failure to produce any evidence on this Frankenstein science of assumptions that is called psychics.

Greatest genius of our time-: Albert Einstein, Thomas Edison (well-funded), or Nikola Tesla (not a Freemason and poorly funded, believed and inspired by the Holy Bible, he was discredited as a result)

The very pillar that holds up modern physics of the study of theoretical physics to form today's doctrine and later taught to all schools and universities was developed by Einstein's theory of relativity. It is also obvious that quantum mechanics is the other pillar that helps form the doctrine of physics itself. He is most known to have created the world's most famous math equation, the mass-energy equivalence formula $E=mc2$. It was predictable that Albert Einstein would also receive the 1921 Nobel Prize in Physics for his services in the field and also for his discovery of the "photoelectric effect," which led to his evolution of quantum theory. After his theory of relativity seemed to be made on hypothetical reasoning at the very best, Nikola Tesla wrote a meme about what he thought of Einstein's theories of physics, that Einstein's theory of relativity was impossible to assert as factual truth and that he thought that his conclusive work on this was "absolute nonsense." Einstein wrote more than 450 papers that contain more than 30,000 documents and many of his thermodynamic fluctuations, statistical physics, and mechanics. Eventually, he proposed a model of atomic matter where each atom oscillates independently. At the same time, quantum mechanics made it possible to discrete the motions of smashing atoms of electrons, which was assembled brilliantly because he applied the Adiabatic principle, which helped when Ernest Rutherford discovered the nucleus and proposed accurately that electrons create a field of high energy and together with Max Plank's energy quanta, which introduced the photon concept. The splitting of the atom was created, releasing a highly charged radiation explosion and was used by the United States on Hiroshima and Nagasaki, Japan, in 1945, that shortly afterward ended World War 2.

Albert Einstein was a Freemason who became a celebrity scientist who continued the heliocentric hoax that predictably is keeping humanity enslaved mentally into subdued ignorance. His theory of relativity was so blatantly purposefully complicated with sophistry that when it was first released to the masses, it was claimed that no more than eight people in the world were capable of understanding it. Public interest was aroused by

Einstein's assumptions of his theories and also by the spectacular manner in which it had been received by the British Royal Astronomical Society on November 6, 1913. The hoopla was that they added a prize of $5,000 if someone could offer the best explanation of what relativity actually is in the form of an essay. The thesis was to describe it so that the public could understand what it was about! Einstein himself quoted, "If you can't explain it simply, you don't understand it well enough." The whole plan was to have it removed from ordinary fact and simple plain English as possible. The physical substance of the theory is elusive, and with ambiguity, locating a concrete testable proposal amounts to finding a way out of an endless labyrinth, as everything is relative to everything else, contained within its own isolated systems that merely "relate" to one another. Relativity is clever, but so is "with-craft." It also belongs in the same predictable category as Sir Isaac Newton's law of gravitation and the Kant-Hershel-Laplace's nebular hypotheses in that it is a superfine effort of the imagination seeking to maintain an impossible theory of a flying ball in defiance of "every fact against it." In Albert Einstein's own words,

> What we mean by relative motion in a general sense is perfectly plain to everyone. If we think of a wagon moving along a street, we know that it is possible to speak of the wagon at rest, and the street in motion, just as well as it is to speak of the wagon in motion and the street at rest. Ignore all of our physical observations. It is a mental exercise whereby we attempt to view events in an isolated void, ignoring their connection and relation to everything else. This is how relativity works; by disconnecting all physical laws and giving them their own "relative playground" where they do not interact and as such cannot contradict each other in their consequences. We are free to view the wagon at rest and the street in motion as long as we ignore everything else around them.

Gerard Hickson, in 1921, in reference to the newly proposed theory of relativity by Albert Einstein:

> It will be remembered how Hipparchus failed to get an angle to the stars 2,000 years ago, and arrived at the

conclusion that they must be infinitely distant; and we have seen how that hypotheses has been handed down to us through all the centuries without question . . . But I can conceive that in the course of time this Relative Phantasmagoria might come to be regarded as science, and be taught as such to the children of the near future.

Throughout the whole of his theories, there are these claims that are also physical, and so he had to create a series of thought experiments intended to show how his principles would work out in general, consisting of leaps of logic over complications and difficulties, which he could not avoid. When he suggested that the street might be moving while the wagon, with its wheels revolving, was standing still, he was asking us to imagine that in a similar manner the Earth we stand upon might be moving while the stars that pass in the night stand still, contrary to all our observations. If you do feel any spinning sensation right now, seek a physician. It is a case of appeal, where Einstein appeals in the name of a convicted Copernican astronomy against the judgement of Michelson-Morley, Nordmeyer, physics, fact, experience, observation, and reason. We, on the other hand, are council for the prosecution, judge, and jury. The first evidence presented against the heliocentric foundation is to be found in the Michelson-Morley experiment of 1887 in Chicago. Professor Michelson, described in the *New York Times* as America's greatest physicist, was foremost in determining the velocity of light and using the speed of light in relation to the rotundity and the velocity of the Earth to determine how fast the Earth was rotating, presumably on its 23.4-degree axis. Astronomers have, for a very long time, stated that the Earth travels around the Sun with a speed of nineteen miles per second, or sixty-six thousand six hundred miles an hour. Without in any way attempting to deny this statement but instead taking it as a given assumption, Michelson and Morley set out to test exactly what was the velocity with which the Earth moved in its orbit around the Sun. A very well-illustrated account of that experiment will be found in the *Sphere*, published in London, July 11, 1921. Here's how it's explained. But to the experimenters' surprise no difference was discernable. The experiment was tried through numerous angles, but the motion through the "ether" was *nil*. This was the best that modern physical science could do to prove the movement of the Earth, and the result *showed that the Earth did not move at all, nil.*

But the world of astronomy "still has not accepted that result," for it continues to preach the old dogma, willing to accept the decisions of their imaginary hypothetical conclusions about Earth traveling through space at one thousand one hundred miles a minute. In a bold distraction, they tried to account for the physicists' failure to discover any movement by finding fault with the ether (or ether). They present the lack of motion as lack of ether with lots of hand waving. So with relativity, Einstein was forced to include several more ideas, each depends upon another and each contributes to support the essence of all his relativity fervor. He has laid practical assumptions and unprovable theories on top of a disproven hypotheses and brilliantly and incorrectly extended a "hoax" and lengthened the playing field of brand-new openings of speculative pseudoscience. Although the Earth has been proven to be stationary and motionless just as our own senses dictate of this reality, Einstein believes that besides spinning and tumbling through space, there "is no ether" and that light is a material thing (is this guy absolutely a looney bird?), which comes to us through empty space. Consequently, light has weight and is therefore subject to the law of gravitation, so that the light coming from a star may bend under its own weight or deviate from the straight line by the mere attraction of the Sun or any other celestial body it has to pass in its journey to the observer much further away than previously supposed.

Every method that was used based on the geometry of Euclid and the triangulation of Hipparchus imaginary conclusions that fail to discover the distance to a 'star.' because its real position is no longer known and can never actually be known for certain. So Einstein has invented a new kind of non-falsifiable geometry to calculate the positions of the stars by what is nothing more or less than metaphysics. Another quote by Einstein:

> The observer is located on the surface of an earth which is rotating on its axis, and at the same time travelling through space at many thousands an hour, consequently his place, or locality, is continually changing with respect to an imaginary point fixed in space. Notwithstanding this change of place, electromagnetic laws appear to act precisely as they would if this place was not changing its position with respect to that point.

So there you have it. After much obfuscation, from the master mason himself, it appears to act precisely as if there was no motion to the Earth whatsoever (Gerard Hickson, 1922). Halley, Newton, Galileo, from Europe to America and all Western civilized academia, to Einstein and then to NASA, astronomy has used the same errors in calculating distance of stars and planets in what is called Astrometry. In no uncertain terms this work completely disproves the "flying ball gravity based heliocentric theory" that has been taught in every classroom on our flat, round, planet Earth for the past five centuries.

Tesla made it his ambition to point out just how wrong that Einstein's theory of relativity actually was and came out rather quickly about it. Tesla always was alone with the inspiring aspect of being able to connect his inventions to the very source that inspired him, the Bible.

Tesla's real only rival of being the world's most brilliant inventor and scientist would be Leonardo Da Vinci, and it's not as debatable as one may suggest. The underfunded Tesla and his mind-boggling inventions were scooped up intentionally from the elitist Freemason Thomas Edison, who in the United States, is better known as inventing electricity and lightbulbs than Tesla, but the chessboard of the internet and the FOIA (The Freedom Of Information Act), along with the fake news mixed in with the misinformation programs from the world's governing body, only go further to put humanity in a static state of ignorance!

I do not and will never go backward in thought ever again, and why should I believe in a false doctrine? Because how can that ever be any credit to one's knowledge that helped continue the doctrine of fantasy space and time and created a math formula but has absolutely zero inventions to his credit? He is considered a genius and absolutely was probably the only man that understood his backward formula to create his theory of relativity. In fact, two of the world's best physicists, at the turn of the nineteenth century, performed an experiment to find out exactly how fast the Earth was spinning or rotating. Albert A. Michelson and Edward Morley performed an experiment and were very well funded to be able to obtain the necessary resources to carry it out. To formulate any authentic experiment and prove the validity of such an endeavor, the Michelson-Morley experiment in 1887 proved something of such significance that almost any veracity of books and documents are in the hands of the NSA and are deemed classified material. I have heard that in 2040 these documents are to be released as public domiciliary information. One of the materials that were used was the speed of light in an ether of space. To facilitate the accuracy of this experiment, they scheduled four different points of embarkation that closely resembled a square. Of course, a square has to actually be drawn anyway to create a perfect spherical circle. The common knowledge of the universe at this time was the evidence that ether exists in space and time. Time and space has been clouded with delusional teachings of an imaginary cosmology that was never based on any evidence of any sort just to propagate a lie over and over again until that lie stops being challenged. At this point, the nihilistic agenda was stopped dead in its tracks. Knowing that ether exists in space as an ascertained verifiable fact, notwithstanding, the Michelson-Morley experiment used the speed of light from a point we deem as A to a point that we deem as B and from point C to point D that formed a perfect square for light to travel and to apply the filters needed to find out exactly how fast the Earth was rotating. Despite the ongoing efforts to clean any mistakes and to try to maintain the integrity of this experiment, Michelson was performing the experiment over and over again to try to make sense of what these findings, and the results were absolutely stunning. The Michelson-Morley finding was that the Earth was not spinning or moving at all. The official count was that it's spinning zero miles per hour. My good God in heaven, these two men were the greatest scientists in the world, and they not only had the resources backed by the elite to conduct an experiment for the ages, but they were able to provide scientifically sound and fundamental evidence of actually physically being able to ascertain a full scientifically accurate experiments on the rotation speed of our Earth. This information and several aspects of this information is the heliocentric globe spinning Luciferian ball Earth death, and it very well should have been, but was it? Not by a long shot.

The world control system elites and the puppets to continue the world's frauds called the Freemasons had to come up with something to save the dying heliocentric imaginary globe. The conclusive experiments of Airy, Michelson, Morley, Gale, Sagnac, Kantor, Nordmeyer, and others were plans to create a dynamic theory of opposition. Of course, let us prove it with unprovable math equations. Albert Einstein created his special theory of relativity, a wicked, nefarious, and brilliant revision of heliocentrism, which, in just a single philosophical swoop, banished the universal ether from scientific study replacing it with a form of relativism, which allowed for heliocentrism and geocentrism to hold equal merit. If there is no absolute etheric medium within which all things can exist, then hypothetically, one can postulate complete relativism with regard to the movement of two objects, such as the Earth and Sun. At the time of this experiment, the Michelson-Morley team was even followed up by another team of Michelson-Gale to reaffirm the veracity of such claims that the Earth does not move! Both of these experiments had long established the existence of ether, but the church of deception, the heliocentric model that needed to be saved by all means, Albert LIEenstein never tried to refute the experiments because he couldn't. You simply cannot unsee the truth, but Einstein snuck around and objected with another notion of "absolute relativity" claiming that all uniform motion is relative and there exists no absolute state of rest anywhere in the universe. Einstein's experiment had to ditch the reality of ether's existence that made his theory work in the model of heliocentric world. It's as though Einstein was suggesting that light bends like how waves bend in the water, but dismissing the ether of space is like creating water waves without the water being the base for the waves to even be real or based on reality. He created a fantasy in the twentieth-century liars of our world ran with it and Einstein was paid an amazing amount of money to live out the rest of his life promoting an image of a "nutty professor" with *zero* inventions to his credit. He was *Time*'s genius of the last one hundred years.

Wow, that is as asinine of a choice that I have ever seen, actually laughing my ass off at this convoluted bullsh*t yet very evil in nature. The heliocentric theory, by putting the Sun at the center of the universe, made man appear to be just one of a possible host of wanderers drifting through a cold godless sky. It also seemed less likely that he was born to live gloriously and to attain paradise upon his death. Finally, contrary to NASA and the modern Masonic astronomical establishment, the Bible reaffirms in several passages that the Earth is motionless.

> He has fixed the earth firm, IMMOVABLE. (1 Chronicles 16:30 and Psalms 96:10)
> God who made the earth and fashioned it, himself fixed it fast. (Isaiah 45:18)
> The world also is established, that it cannot be moved. (Psalms 93:1)

Nikola Tesla, the man who invented the twentieth century!

Nikola Tesla was a Serbian American inventor, engineer, and physicist. He is documented and well known for his invention of the modern generation of alternating current electrical power. He actually has become even more thought of than his contemporaries Marie Curie and Albert Einstein in the academia mainstream of today, and the continued effort to find and then release information that has been suppressed from benefiting the good of humanity is possible today but sometimes still very hard to obtain unless you are looking for it through independent means. Tesla emigrated to the United States in 1884 to work with Freemason Thomas Edison, who, by the way, was very well-financed by the secret society and the Masons. Tesla always studied the Bible every day and every night, and it was this inspiration that brought him pretty solid local backing of his many projects but never the financial backing from the Freemasons, who literally carry the keys to all the bank vaults in the world. By 1891, Tesla's inventions included the system of arc lighting that was truly inspired by Moses when he received the arc of the covenant from God. He also invented the internal link-alternating current motor-power generation and transmission systems, and that was in 1888. In essence, he invented the cellular phones almost 130 years ago. He also invented the systems of electrical conversion and distribution by Internal link oscillatory discharges. By 1891, he further invented the wireless transmission of electrical power called the Tesla coil transformer. Tesla always said, "He who controls the photons, controls the universe." On the sequence of how Tesla opened up his mind to grasp not only the holy book of God but also how it manifested to him being blessed with the very best inventor probably in the history of mankind or at least the last five hundred years. The only man who can rival Tesla undoubtedly is Leonardo Da Vinci, who incidentally invented the helicopter. In Tesla's many readings that drew up and ramped up his inspiration, one can go to the book of Genesis, and when "God created" on the

> First day: Let there be light (photons, light)
> Second day: Hydrogen (water in gas form)
> Third day: Elements (earth and plants)
> Fourth day: Cosmic bodies
> Fifth day: Creatures of water (fish)
> Sixth day: Creatures of the land (animals and man)

Sadly, in the end, Tesla died alone in a New York Hotel due to coronary thrombosis. It has only been pretty recent that the work from him has been truly transparent for all to resonate and understand that his achievements in the field of science is truly remarkable. Some writings about Tesla was posted on October 10, 2015, by Michael Thomas:

> Be alone, that is the secret of invention; be alone, that is when ideas are born. (Tesla)
>
> Let the future tell the truth, and evaluate each one according to his work and accomplishments. The present is theirs; the future, for which I have really worked, is mine. (Tesla, April 1927)
>
> In the twenty-first century, the robot will take the place which slave labor occupied in ancient civilization. (Tesla, February1937)
>
> Fights between individuals, as well as governments and nations, invariably result from misunderstandings in the broadcast interpretation of this term. Misunderstandings are always caused by the inability of appreciating one another's point of view. (Tesla, January 7, 1905)
>
> The scientific man does not aim at an immediate result. He does not expect that his advanced ideas will be readily taken up. His work is like that of the planter-for the future. His duty is to lay the foundation for those who are to come, and point the way. He lives and labors and hopes. (Tesla, June1900)
>
> Peace can only come as a natural consequence of universal enlightenment." (Tesla, 1919)
>
> It is paradoxical, yet true, to say, that the more we know, the more ignorant we become in the absolute sense, for it is only through enlightenment that we become conscious of our limitations Precisely one of the most gratifying results of intellectual evolution is the continuous opening up of new and greater prospects. (Tesla, September 9, 1915)
>
> Invention is the most important product of man's creative brain. The ultimate purpose is the complete mastery of mind over the material world, the harnessing of human nature to human needs. (Tesla, 1919)

Throughout space there is energy. Is this energy static or kinetic! If static our hopes are in vain; if kinetic-and this we know it is, for certain-then it is a mere question of time when men will succeed in attaching their machinery to the wheelwork of nature. (Tesla, February 1892)

The progressive development of man is vitally dependent on invention. (Tesla, 1919)

Life is and will ever remain an equation incapable of solution, but it contains certain known factors. (Tesla, 1937)

The day science begins to study non-physical phenomena, it will make more progress in one decade than in all the previous centuries of its existence. (Tesla, unknown date)

The desire that guides me in all I do is the desire to harness the forces of nature to the service of mankind. (Tesla, July1934)

Every living being is an engine geared to the wheelwork of the universe. Through seemingly affected only by its immediate surrounding, the sphere of external influence extends to infinite distance. (Tesla, February 7, 1915)

The last twenty-nine days of the month are the toughest! (Tesla, 1919)

The individual is ephemeral, races and nations come and pass away, but man remains. Therein lies the profound difference between the individual and the whole. (Tesla, June 1900)

Though free to think and act, we are held together, like the stars in the "firmament," with ties inseparable. These ties cannot be seen, but we can feel them. (Tesla, June 1900)

I do not think there is any thrill than can go through the human heart like that felt by the inventor as he sees some creation of the brain unfolding to success . . . such emotions make a man forget food, sleep, friends, love, everything. (Tesla, June 7, 1896)

The spread of civilization may be likened to a fire; first, a feeble spark, next a flickering flame, then a mighty blaze, ever increasing in speed and power. (Tesla, January 16, 1910)

Of all the frictional resistances, the one that most retards human movement is ignorance, what Buddha called the greatest evil in the world. (Tesla, June 1900)

Our senses enable us to perceive only a minute portion of the outside world. (Tesla, January 7, 1905)

Three possible solutions of the great problem of increasing Human Energy are answered by the three words; food, peace, work. (Tesla, June 1900)

Our virtues and our failings are inseparable, like force and matter. When they separate, man is no more. (Tesla, June 1900)

Money does not represent such a value as men have placed upon it. All my money has been invested into experiments with which I have made new discoveries enabling mankind to have a little easier life. (Tesla, April 1927)

This planet, with all its appalling immensity, is to electric currents virtually no more than a small metal ball. (Tesla, March 5, 1904)

Instinct is something which transcends knowledge. We have, undoubtedly, certain finer fibers that enable us to perceive truths when logical deduction, or any other willful effort of the brain, is futile. (Tesla, 1919)

The scientists of today think deeply instead of clearly. One must be sane to think clearly, but one can think deeply and be quite insane. (Tesla, July 1934)

Today's scientists have substituted mathematics for experiments, and they wander off through equation after equation, and eventually build a structure which has no relation to reality. (Tesla, July 1934)

The GLOBE AND THE ARROGANCE OF OUR INSIGNIFICANCE OF OUR ORIGINS- IS THE MASTER OF DECEPTIONS ATTEMPT TO DIVORCE MANKIND FROM HIS CREATOR- THE BATTLE FOR YOUR SOUL

Curvature- Is impossible, it merely defies the very basic law of psychics. Water cannot curve, and does not ever curve from any convexity that can honestly be ascertained to exist at any level of being a sound person that has the ability to do very simple real science experiments, and more people have and are doing these kind of experiments and are waking up from the deep sleep that over 80% of the world is still "sawing Logs and continue to be in a trance of being further molested into a false doctrine of utter non-sense. The fact that surveyors and engineers and Architects never follow the supposed given allowance on the curvature of the earth based on a 25,000 circumference, on the spherical trigonometry Chart is by any of these guys who laid down the original massive transportation systems throughout the United States are all dead now, but the fingerprints of truth are so evident everywhere that you look, but the exact measurements and calculations on 100% accuracy for all of these projects are then, as they are now, based on a flat earth horizon with absolutely zero convexity of defying the very nature of water, and the laws of psychics. Water will always find its level, its flat level, and it cannot curve upwards nor downwards. The fact that just 34 degrees south Latitude in Australia, we are being told that up is down and down is up. Meanwhile the people in Australia are all standing on their respective heads, they are upside down. We are told that the earth is spinning but your very senses are usually right almost always, and again, the lie is the mother of all lies. It is the most precious lie and the most significant disclosure that will ultimately be freed from the spiritual slavery that these atheistic monsters have mostly been able to successfully do.

I am very open minded person, and will forever be a sponge to listen to any human being whose seen the curvature of our earth. I mean if it absolutely existed, i also would assume that it would be a tremendous tourist attraction to actually see the exact coordinates of where this damn curvature exists, because according to the Global elite Liars and Frauds very own map and measurements on their own calculations of earth spinning ball illusion that our kindergarten teacher all taught us as a 5 year old child, the Pythagorean table of spherical trigonometry tells us approximately 25,000 statute miles in equatorial circumference with a curvature of 7.935 inches to the mile, varying inversely as the square of the distance, meaning in 3 miles there is a declination of almost 6 feet, and what really pisses me off, is that on the worlds longest bridge over water, a bridge in China is 106 miles long that has to account for over one mile for curvature of the earth, yet, the bridge makers, and the engineers, and stripping it down from the blueprints to it's completion, there was no such allowances for over 1 mile of curvature? Are you lying to me, I ask myself? No way, it's a fact of life that people are living in a star wars illusion, and the awakening of what our reality actually is all about, is very frighteningly hard for anybody to ever swallow, because, we can even attempt to just snap out of the lies and fantasy theories of who we are, what we are, where did we come from, who made us, and the poisonous indoctrination of false education, and as a youth the hard work started to become as educated into the system as one could possibly be, and the deeper and the brighter that we excelled at fraudulent information that we took as being truth, the more older and more into the pseudo sciences, and making scientific mathematical equations to help determine what can be ascertained as fact, especially making a career out of this misled and misguided education that ends all the way through the very fabric of all university institutions all over the plane of earth. The masters degrees earned that almost always unbeknownst to our hard working young men and women, who sacrificed a ton of fun and folly to earn their way to the highest levels of that specialty in science that he or she earned. It is hard to let go of your ego, and the years of false education that dominates ones persona. The ones who are so deep into the indoctrinated system, whom I affectionately call the sheep, I pray with all my might, that the walls of secrecy concerning the worlds oldest secret society are coming down, but it will come at a very very high price for all of humanity.

I will list below a very compelling argument that the very senses that God gave you when you're observing the horizon, are correct always. The field of vision from the observer will absolutely converge to the level of your 2 eyes, period! Its insanely simple, but I'm sure the globetards may figure another back-ass-backwards Mathematical formula to first dilute reasoning, and convince you that we are spinning 1,068 miles per hour on a spinning globe, yet your own god given sense is telling you that you're not moving at all. Shit, when you and I were toddlers and on a tri-cycle going 2-3 mph-you knew then that you can feel yourself moving, and they are also suggesting that we are racing around our sun at 68,000 mph, but yet, we look up to the sky and all of the constellations are exactly the same for thousands of years, so what you are seeing, or what you're not feeling, is all incorrectly wrong, because the Free Masons like Neil Degrasse Tyson says so!! You saw it on TV, or science told you so, or you looked into on your idiotic wikipedia (US GOVERNMENT SHILL CONTROLLED) Cell device told you that world is a globe. It is astounding how much lack of cognitive thinking, and deeper or critical thinkers, that I do not see!! I see a world of sleepers, people who are never going to break out of the grand deception. Below is a compilation of absolute proofs that we have zero curvature that matches a real reality.

1. Lighthouses and countless historical observations from Surveyors, Engineers, "The lights which are exhibited in lighthouses are seen by navigators at distances at which, according to the scale of

the supposed curvature that is given by our friendly free mason astronomers, they ought to be many hundreds of feet, in some cases, down below the line of sight! For instance:

Cape Hatteras is seen at such a distance (40 miles) that, according to theory, it ought to be nine-hundred feet higher above the level of the sea than it absolutely is, in order to be visible! This is a conclusive proof that there is "NO CURVATURE" on the surface of the sea- the level of the sea',- ridiculous though it is to be under the necessity of proving it at all: but it is, nevertheless, a conclusive proof that the earth is not a fricking Globe."

2. The Isle of Wright Lighthouse in England is 180 feet high and can be seen up to 42 miles away, a distance at which modern astronomers say the light should absolutely fall 996 Feet below line of sight. The Cape L'Agulhas Lighthouse in South Africa is 33 feet tall, 238 feet above sea level, and can be seen for over 50 miles. If the world was a globe, this light would fall an almost incredible 1,400 feet below an observer's line of sight !!

3. The Statue of Liberty-New York, stands 326 feet above sea level and on a clear day can be seen from 60 miles away. if the earth was a globe, that would put our symbol of liberty lady at an impossible 2,074 feet below the horizon!

4. The Lighthouse at Port Said-Egypt-has an elevation of only about 59 feet, but actually has been seen 58 long, long miles away. where according to the pseudo taught astronomy that are taught by astronomers, it should be almost a half -mile (2,074) feet below the horizon!

5. Engineer- W. Winckler- wrote in a magazine called Earth Review October, 1893 when asked about the supposed allowance of 8in. per mile squared from Spherical Trigonometry, suggested instead that regular trig. is used based on a flat horizontal line irregardless of any other measurements, stated, "As an Engineer of many years standing, I saw that this absurd allowance is only permitted in school books. No engineer would dream of allowing anything of the kind. I have projected many miles of railways and many

more of canals and the allowance has not even been thought of, much less allowed for. This allowance for curvature means this- that it is 8" for the first mile of the canal, and increasing at the ratio by the square of the distance in miles; thus a small navigable canal for boats, say 30 miles long, will have, by the above rule an allowance for curvature of 600 feet. Think of that and then please credit engineers as not being quite such fools. Nothing of the sort is allowed. We no more think of allowing 600 feet for a line of 30 miles of railway or canal, then of wasting our time trying to square the circle"

6. Suez Canal-connects the Mediterranean Sea with the Gulf of Suez on the Red Sea- The canal is 100 miles long-It was constructed without any allowance for the earths curvature being considered. They dug along a horizontal datum line 26 ft. below sea level, passing through several lakes from one sea to the other sea 100 miles away. The amazing disparity of all of this, is that the average level of the Mediterranean is 1/2 foot above the Red Sea, and the floodtides in the Red Sea rise 4 feet above the highest and drop 3 feet below the lowest in the Mediterranean making the half-tide level of the Red Sea, and the 100 miles of water in the canal, a continuation of the same exact horizontal line !! The curved line in our Globe, according to the Free Mason conscience, or sub- conscious of a 25,000 mi. spherical globe very own calculations especially applying the correct math to all measurements, the water in the center of the canal would be an unbelievable 1,666 feet (50 squared x 8 inches = 1666 feet, 8 inches.

David Wardlaw Scott-author of a book called "Terra Firma"- (page 134)

"The distance between the Red Sea at Suez and the Mediterranean Sea is 100 miles, the datum line of the canal being 26 feet below the level of the Mediterranean, and is continued horizontally the whole way from sea to sea, there not being a single lock on the canal, the surface of the water being parallel with the datum line. It is thus clear that there is no curvature or globularity for the whole 100 miles between both seas; had there been, according to the Astronomic theory, the middle of the canal would have been 1,666 feet higher then at either end, whereas the canal is perfectly horizontal for the whole entire distance.

The Great Canal of China, said to be 700 miles in length, was made without regard to any allowance for curvature, as the Chinese believe the Earth to be a Stationary Plane. I may also add that no allowance was made for it in the North Sea Canal, or the Manchester Ship Canal, both recently constructed, thus clearly proving that there is no globularity in Earth or Sea, so that the world cannot possibly be a Planet."

HOW CAN RIVERS RUN UPHILL-Water will always find it's Level. The very principles of psychics which is common knowledge that can be verified without any interference of having to create convoluted scenario's of a theoretical analysis to hide facts, or to cover-up virtual truth, which at this point in time, has to be regarded as pseudo-science. I assume that if I can just go to the magical place where standing water is curving on an inclination that can confirm the magical GRAVITY GLUE-that holds all the waters in the ocean on our spinning ball earth, and yet it cannot keep a bird from flying or keep insects from flying, all the butterflies, and yet, just looking at a globe that I embarrassingly purchased at a Barnes N Noble right around the Christmas Holiday's in 2015. I looked at several areas in Africa where the Congo river would have to run upstream for better then 900 miles, which is only possible in Hollywood movies, and false theories that were created by The Global Elites, in particular, the Free Masons,who were the Knights Templar's- the foot soldiers for the Jesuits in Rome, Italy. After all, it was essentially us who killed Jesus Christ. The Jews and the Romans, and the paradigm shift from Geo-centrism to the Heliocentric model provided the divorce of human kind from significance that was made in the very image of God himself, and transformed the essence of purity of ones soul and the very fact that to much attention is always paid the very to a spec of insignificance.

On October 14, 2012, Austrian Skydiver and daredevil Felix Baumgartner jumped from a helium balloon in the stratosphere from a height of 128,000 feet. It was shown from a Red-Bull commercial advertisement and of course the main course was not necessarily the jump as much as it was of showing a 180 degree curvature of the earth as a global ball. Unfortunately for Mr. Baumgartner, and the carelessness of the flight crew during commercial footage as he was preparing for his historic jump, If one looks closely out the window of his small cabin, you can certainly see the beautiful completely flat horizon as final preparations and instructions were carefully coordinated for his historic jump to set a standard world record that will be very hard to replicate, at least anytime soon. It is quite interesting that many people worldwide have sent weather balloons as high up in the stratosphere at 126,000 feet with a Dog -Cam taking electrifying footage of the flat earth plain and, I do mean a complete straight line as far as your eyes can possibly see . Eventually, the horizon is always at the level of the dog-cam camera which is absolutely impossible if we were on a globe. If anyone simply picks up a basketball and to put a visual aid to enhance what would happen in explaining this with a prop such as a basketball, you can attest to the fact that on a ball-earth on a 25,000 circumference of how much earth is actually underneath our feet, you will have to agree that the horizon would drop and descend away from the level of your eyesight, not the other way around.

It simply is almost too tough to understand that we as a society have to be so reliant and gullible on accepting what someone taught them in school, or the regurgitating of the globe being shoved down the throat of all the masses for so very long, that its almost beyond comprehension that all of these proofs and re-proofs of the flat earth truth, and simply not accepting the brainwashing of what is real and what's not real that was part of the indoctrination of what I call pseudo science, or has no better value then what amounts to "science fiction". I call an ace of spade for what it is, and I can't seem to understand why in modern science that was taught to me as a child, were only based on theory, assumptions, and equatorial math to help moderate the message and to further implement a doctrine that has become poison for not just humanity itself, but the most venomous poison that dulls the senses and leaks out the very real spiritual connection that was already part of the human biological make-up that provides various abilities for human interaction with his thoughts and the inner -self of ones real ability to look at all avenues of research. The inability of people that are not able to get out of this prison matrix of significant insignificance, has put humanity at the edge of the cliff of becoming so dull to what their sense of purpose, the sense of worth, the sense of being able to connect your physical being to the metaphysical and most of all, to be able to connect to the spiritual realm.The Agenda from the WCS-World Control System has been in place since before the

American Revolution and the official birth of the United States of America that actually took place. The very inordinate plan of dominant domain started on May 1st, 1776, and the unconventional opinion was that the British Empire had indeed become the most powerful world power in the world. The fact that Albert Pike was the George Soros of the late 1700's in that the instigation's of the American Revolution that ultimately led to the war between the Americans and the British, but the war was serving then as it is now, a major fuse of instability had to be planned and orchestrated by the European led Cetral Jewish Bankers who created and implemented such a plan and Free Mason Albert Pike-who represented the interests of profiting financially in a very complete and magnificent way as a means to continue to amass a fortune that especially today, can never be matched in any significant way. It's only because of the 1st Amendment protections under U.S. Law and the age of the internet, and the fact that more people then ever before do not trust what their government is telling them. The amazing thing is that the very word definition of Government, The word GOVERN-From old Greek Word: To guide. The word MENT:From a Latin word meaning: Mind. Government meaning is quite simply, " TO GUIDE YOUR MIND". It was only 85 years later, in 1861, when the Civil War, the WAR BETWEEN THE STATES, did the World Central Bank's very powerful Jewish family owned banks, who then used their unmatched wealth and unmatched resources to deliver the absolute knockout blow that officially occurred in 1914 and signed into law by President Woodrow Wilson, who is by all means, in my belief, to be the very worst President in the history of the United States. The Civil war in 1861 was instigated by the WCB in a unilateral move to control the very wealth of America's money. On one side, you have the patriot warriors of a sovereign state that will naturally favor the U.S. Federal Reserve Bank, against the 13 Banking Jewish European Families whose quest of who will control America's wealth, that absolutely led ultimately to a blue-print plan to use as a complete takeover of an unanswerable question of who started the war, which will be then, as it is now, about the almighty dollar. This detrmined stance would have to have all the pieces in place, in order to master the plan, and thus carry it out to it's full realm, and to create caos for other violence, and to procreate, or just simply cause consternation and disdain between the very peoples of the United States that would propel the American Civil War in 1861 and the explosive devices of unrest that lead to hate, divisions between people and goods and services, and transports between neigboring states, and the tragic events played out with millions upon millions of deaths in the United States between the North and the South, which was divided essentially by the Mason Dixon Line, which was set up by Charles Mason in 1763, and then again with Jeremiah Dixon in 1767. The Jewish colloboration with the Roman Empire's gift to humanity, The Vatican in Rome, which with the money assembled by the World Central Bankers even during the French Revolution when Napoleon was in his glory, and the perfect components were in place as the catalyst of war was near, the gunpowder, or the tilting keg to amplify the impact of an explosion of no return. In other words, the Civil War most definitely had a backdrop of the South using African Americans as slaves that was very cruel and harsh irregardless of any excuse borne by any man to defend on any level of reasoning. The irony of a lot of what happened in that time, was that a lot of the special privileged families of the North, were also in the businesss of being slave owners themselves, including the Free Mason Presidents of our beloved country. But, the main characters of why the Civil War

PART 2---NASA- "MARS" and all of the Rovers do their tric discsks and fake scientific experiments not on the wandering star called "MARS" in which their simply isn't anything to even land anything on, and the masses and some deceived with the continued research on something that has no bodily form of any solid matter, they are not opaque masses of matter, as it is believed, they are simply immaterial, luminous and transparent discs. The Sun is not 93,000,000 million miles away, and crepuscular rays' are quite evident that the sun is probably 3500 miles away and the psychopathic pseudo science of theoretical psychics used calculation of mathematical formulas to try to prove something that is quite intentionally wrong minded because it was thus taught as reality to children to further the dialogue of our originality of our standing of human consciousness of spirit and of flesh. The more educated one receives from this utter non-sense, the further apart his natural spirit conscious was turned away from reliable consensus, to think for one-self! Part of "faking space" is certainly only attainable through changing the origins of the "fallen angels" who magically appeared from the heavens, and it's origins no doubt come from the Book of Enoch, but has manifested man to change the characters of its inter-dimensional factual reality to a different narrative altogether. "You simply can't re-brand Fallen Angels to Extraterrestrials without the Globe Model !"-Mark S Hollander

You and I will see a different view of what Mars, Jupiter, Saturn, Venus, Neptune, Mercury, Pluto, and Uranus, through an un-digitized telescope and all of the biblical seven wandering stars which these lying maggots of deception actually create artists renditions to all of these so called imaginary planets and a lot of humanity are convinced that these fake CGI -Photoshop digital graphics are real, and it is up to all of us to spread the message of this gigantic deception that has emanated only a tremendous disconnect to what is real in our lives today, we do see what is actually fake, and we see it every single image that NASA puts out about space and it's space imagery. When I point out the impossibility of even being able to descend on anything on the scope and the projections that NASA has claimed, it should be cause for a major consciousness shift to many who will take the time to do research on what is stated throughout this project. Even assuming Mars was an actual spherical desert planet as NASA claims, it is quite impossible for them to have safely landed the probes based on their own trials and statistics. They claim that the surface pressure on Mars is only 3/10ths of 1% the surface pressure on Earth, and equivalent to the pressure at about 23 miles above the Earth. There simply is not enough air -matter at that pressure to provide any lift for opening and billowing out the parachutes NASA to land its Mars probes. There has never been a parachute that has ever been designed that has ever been able to deploy a parachute successfully at that ridiculous altitude. I remember Joe Kittenger's record highest, fastest, and longest parachute dive from the Earth's upper atmosphere had him free-falling from only 19 miles up for 15 minutes at an incredible 767 mph and his drogue chute proved useless and offered no deceleration. Yet NASA would have us believe, for example, that Phoenix's parachute managed to somehow slow it down from 12,738 mph to 123 mph in just 2.86 minutes before its final landing. In other words, NASA is claiming to do something on Mars that we HAVE ZERO EVIDENCE THAT IS EVEN POSSIBLE TO DO ON EARTH AT A SIGNIFICANTLY LOWER ALTITUDE AND AT 16 TIMES SLOWER SPEED!!!

 I want to get to a more elaborate evaluation on the continue faking space hoax that continues to entertain us through trickery, wizardry, and have taken our imaginations into the cosmos, seemingly never to return! Another astounding collaborative unity between NASA and Hollywood- which actually is part of the third partner in spinning the Heliocentric deception at all costs and at every single level to keep everybody in the dark about what is truth and the importance of mankind with the weight of having his soul ingested with the atheistic system that absolutely disconnected our spiritual awareness of our true destiny. It doesn't belong to the cosmos, it belongs to the creator of the "heaven's and the Earth, God our holy father who reigns above the separated waters as has been foretold in Genesis.Lets examine the impossibility of space travel and base this on 2 different versions of the Mars lander making a successful landing on the surface of Mars.

"On July 14, 1976 the orbiter modual which weighed 5,125 pounds detached its lander. I can find no listed weight in any encyclopedia on space but since it could carry up to 638 lbs. of fuel in addition to its payload that Lander had to weigh at least 1,000 lbs. NASA claims that after the lander was detached rockets were used to slow it down to 560 mph at an altitude of 800,000 feet. Then it was allowed to fall 781,000 feet under Martian gravity before a parachute was deployed at 19,000 feet. At 4,600 feet this chute was detached and NASA tells us the velocity went down to 145 mph. Rocket engines under computer control then landed it. Martian gravity is about .37 that of Earth. Earth's gravity accelerates an object at 32 feet per second. This gives Mars the ability to accelerate an object at 11.84 feet per second. The 560 miles per hour horizontal motion will not affect the downward velocity of an object that falls 781,000 feet on Mars. The terminal velocity at the time the chute was deployed was about 4,300 feet per second which is almost 3,000 miles per hour. That's much faster than a speeding bullet. NASA claims that in a matter of 14,400 feet that chute operating at or under near vacuum conditions reduced the Lander's speed to 145 miles per hour. Sure it did ! That was then, let's look now. The next probe to land on Mars did so on July 4, 1997. NASA tells us that the "Pathfinder" came in at 16,600 miles per hour and was jettisoned to boldly plunge into the fringes of the Martian atmosphere without using retrorockets to enter orbit. As usual, there were two different histories given by NASA. The first states by by some miracle during the next minute it's speed reduced to 1,000 miles per hour. The second states thatit was jettisoned at 5,300 miles and it's speed was reduced in 30 minutes while it fell to 80 miles. In the second case the Pathfinder would be at the 80 mile high place still doing 4,280 miles per hour. The NASA story gets murky, but it is assumed the Pathfinder was again allowed to free fall until it was 7 miles high when NASA claims the parachute opened. Instead of streaming because it had been popped in almost a vacuum, it billowed forth and slowed the Pathfinder down. 'When it was one mile up it dropped the chute, blew up the airbag, and fired retrorockets reducing its speed to 23 miles per hour. Then the air bag hit the ground and bounced either 3 times or 16 times' [depending on which official NASA source you ' believe'] Ralph Rene, "NASA Mooned America !"

In 2001, an award filmaker by the name of Bart Sibrell produced a documentary called, "A Funny Thing Happened on the way to the Moon." When Mr. Sibrell asked NASA officials to send him footage for his documentary, he was sent an official raw NASA movie clip by a whistleblower from them or it was by mistake, nonetheless, the official NASA clip shows the three Apollo Astronauts Buzz Aldrin, Neil Armstrong,and Michael Collins, for about forty five minutes these astronauts were discussing camera tricks and using transparencies, and props to fake shots of a round Earth ! While all of this was happening they were in communication on audio with Mission Control in Houston, and were discussing how to accurately stage the shot. Someone else was prompting them on how to effectively manipulate the camera to achieve the most desired effect. They blacked out all the windows except for a downward facing circular one, which they aimed the camera towards from several feet away. This created the illusion of a ball-shaped Earth surrounded by the blackness of space. The whole prompt was just a round window in their dark cabin. Neil Armstrong chimed on the audio that he was 130,000 miles from Earth, half-way to the Moon, but when the camera-tricks were finished the viewer could see for themselves the astronauts were not more then a hundred miles above the Earth's surface.

> "Many gullible people still accept NASA's claim of sending men to the Moon, without bothering to carry out any research, or investigation, to see if NASA are indeed telling the truth. There are some who will NEVER accept the Moon missions were faked, regardless of how much factual evidence of a fake is put before them. "- Sam Colby,NASA- " Numerous Anomalies and Scams Abound"

As I get to a second documentary by Sibrell in 2004 that was entitled "Astronauts Gone Wild" where he set out to film the astronauts while asking several of them "to swear on the Bible that they walked on the Moon. John Young from Apollo 10 and 16 threatened to " knock him in the head" then heartlessly ran away to a waiting elevator. Ed Mitchell kicked in the ass with his foot and threatened to kill him by shooting him ! Buzz Aldrin was caught on camera punching Bart Sibrell in the face ! The second piece of his work is quite telling on the very psychological make-up of these ex-astronauts who are caught in this film repeatedly squirming and the vulgarity starts that leads to a few threats of violence from folks like Michael Collins telling them to get the fucking camera out of my face, turn it off! " These Freemasons behave more like pathological liars than honorable cosmonauts. Many have battled with alcoholism and severe depression, as Neil Armstrong apparently did. The sixth astronaut who walked on the Moon's surface, at least that's what the Freemasons' are saying, and are we to believe them because they have been caught with such lies right on their own channels, and the answer to that is they have succeeded at masquerading as magicians in what amounts to being only extraordinary computer graphics and digital technology and have convinced most of humanity to believing into any of these at even face value to being real, and all space agencies of the world should be forgiven because most of the employers that work at these are very departmentalized and little outside that bubble department that you're employed with, will you ever get any substance of what the other department was working on or what they are trying to develop, and the logistical makeup was blue-printed from NASA in the fall of 1958, is this extraordinarily ambitiously space agency was founded. As June closes out in a boiling New Orleans mid summer day turns into evening, I noticed the sun going down and the other end of the sky a fully illuminated light of a full moon and it had absolutely nothing to do with anything the sun was giving out especially at 7:07pm, and the trained eye for a truth seeker can clearly see that there's no parallax to the sun's diminishing light going down on the horizon and everything about this magnificent and so awe-inspiring beauty allows me to bask just how beautiful that the very sculptor of this is the hand of God.' But my observation of our sun going down as the sun itself starts to vanish away from the variance and perspective of our vision. I hate to break it to people who simply do not realize that"you cannot see forever! " When Devon Island was chosen to build a first -class NASA Facility and the fact that all of these animals have been filmed by both landers.

    I do want to cover our Rovers on "Mars" real live testing grounds on Earth that has more than just a few of us curious to what our Rovers doing at a desolate North Canadian uninhabited island that has an eerily reddish landscape that would be a great testing grounds to create real life-like images of Mars but the only problem is that several animals, including rodents and squirrels have been caught in the Curiosity and the Opportunity Rovers that are purportedly on the Mars Landscape itself.

    Devon Island is a northeast Canadian Island that is uninhabited and a surreal landscape that you could absolutely think was on the Martian landscape that NASA and science fiction Hollywood have done equally great, especially at creating a real likeness to a scene that looks real and makes the audience feel the emotional sensation of a great actor, or a great tune and nobody does it better then the Wizards of Westwood, and I am targeting Hollywood also because they are partners to the Freemason -Elites of the 33rd kind, to rewrite would is real and to create a new aura through technology and by nihilistic means to place humanity cosmic poisoned consciousness into a new artificial yet false reality.

The Northeastern Arctic Region Canadian Island is the world's biggest uninhabited island. I am not alone to suggest that an increasing number of people who believe that NASA's Rovers never left the Earth in the first place and the pictures they are allegedly sending back from Mars are taken in remote areas of our planet.

If NASA 's Rovers are not on Mars, one of the biblical seven wandering stars that is another lie to propagate a spheroid-planet of spinning balls in space. Then where are they, then? The fact of it all is that NASA has a full-Laboratory complex on Devon Island, and the very first red-flag for another magical elaborate hoax by the Lucifereans, but it gets a lot better folks! NASA is playing around with these small rovers at Devon Island and this a fact that is well documented and has been officially confirmed by NASA officials that that these practices are filmed by NASA as well as others. Another popular Earth location to fake the Martian Landscape us in the Mojave Desert, and some locals have seen some very odd things going on in the desert. Devon Island is uninhabited and has strikingly similar characteristics as what NASA tells us through tinting the contrast in the color cycle to more reddish tint that they show us in their fake paintings of the seven wandering stars, and of course Mars exquisite paintings takes precision by the fake spherical design to make Mars into something that it simply is not, a planet! Mars is a luminous light that is on a elliptical orbit that the other fixed stars in the sky are not, and so are Jupiter, Venus, Mercury, Saturn, Neptune, Uranus, but planetary bodies they are most certainly not. The only telescopes that see the planets the way NASA depicts them to the hoodwinked masses called sleeping sheeple, the people who are hooked on something that has no provability to exist, space. Since NASA deemed Mars the red planet and what better place where fake Martian photos could be taken? Of course to create the Hollywood-NASA-Disney reality since I deem these 3 companies to be more in reality with Curly, Larry, and Moe. The Three Stooges are fairy tale land ! Nasa uses color filters like the tint on your TV to create their fictional red planet landscape. This full blown deception isn't too big of a secret, and just like the fake spacewalks are all simulated in the world's biggest built in swimming pool inside Johnson Space Center headquarters. NASA does admit to the color tampering of the Mars photos, and we all have been duped into believing the "myth that Mars is the red planet !" This is so easily debunked with any high powered non-digital telescope, because the digital telescopes are pre-fitted with seven fake digital images that is stored into the telescope to further delusion Alize the visual observer. Mars isn't red at all, it looks exactly like how it's described biblically, and it certainly doesn't look anything that NASA shows in their fake images and paintings of an imaginary world of Disney and pass them off as a real photo through computer generated artificial images of these so called planetary bodies. It looks like a row of lights without anything else almost like a twinkling Christmas light in the sky, and even when you're looking at Venus or any of the others, they all look like luminous lights that zig- zag and a dotted middle as if to suggest that it is part of the ceiling decoration of our creator. The ceiling acts as a chandelier moving like a merry go round of translucent and bright lights of no evidence at all of being able to land on the Moon, much less land a rover on Mars. Mars is not a planetary body that any man or woman could possibly land on, much less some toys from Mattel, and I'm being facetious, but, please stop lying to the American people and the other space agencies like Russia and China to stop lying to their people about this, because you have installed a false realism and a real false reality and its a matrix that is a prison, and you have stupefied people senseless with your lies and deceptions of what our very reality is, and what it is not ! THERE IS NO SUCH THING AS SPACE-It is a provable fake and any research person need not be a researcher, per se, you just have to start uncovering the lies one at a time, just the more you uncover, the more broader and widespread lies that you will start to see and revealing all of their hidden closets of evil and of evil intent, the globe-elites and their forbidden secrets to keep our spiritual maturation from ever realizing what it's actually supposed to be used for. Brain power is when you unite your spirit and your flesh simultaneously and it opens up your mind and the unused portions of your noggin will be unlocked for the rest of your days, that's one promise that I can keep! NASA does admit on numerous occasions to alter the colors of their pictures before posting them online, so that they look more like what NASA has suggested and lied about.

"Getting the colors right is not an exact science, says on official Mr. Bell. "Giving am approximate view of what we'd see if we were there involves an Artistic visionary element as well-after all no ones ever been there before." However, great pains are taken to be as accurate as possible, short of going there ourselves.To give people a sense of being on Mars scientists combine views through telescopes data from past Mars missions, and new information from the current mission to create a color-balanced uniform scene.Color-corrected Mosaics simulate the view a person would see if all images in the mosaic were taken on the same day, at the same moment. In addition, the rovers can take three pictures in a row of the same surface terrain on the Mars surface using three primary color filters red, green, and blue to make one color image. " It works like an inkjet printer which combines primary colors to create various shades on paper." explains Eric DeJong, lead for the Solar system Visualization Team at the Jet Propulsion Laboratory. "then,we can tweak the color just like you can adjust the color balance on a TV screen at home." NASA and JPL Mars actually in real reality is a light dark brown, sort of what Earth would look like without water and vegetation. Mars atmosphere is going to be the exact same as ours because its not nearly as far away as we are being told, and what we were taught. From the time that the astronomer Giovanni Cassini-yet another 33rd Freemason member was another pioneer of deception and suppressing truth that has affected our real-state of being.

From a text message that I sent to some of my family members on Thursday, June 26th, 2017-Mark S Hollander-from a hero of our movement of truth,Jeranism.

"Think about this!! If you took a 747 Jet airliner and went to "Mars"-going 500 miles per hour, it would take you 32 years just to get to one of the biblical wandering stars called Mars ! Just to get there, if it's really that far away which is B.S. Think about it, 64 years there and back ! "-That is the definition of "FRANKENSTEIN" space logic.

Some anomalies that have accidentally occurred during the Rovers pictures were done with a stigmatized yet memorable with the "MARS RAT" being at the epic center of possibly the biggest farce since the moon landings where it looms that a whistle-blower from NASA that intentionally kept the photos, they determined that the Mars-rat was in fact, an Arctic Lemming which is commonly found near the Arctic circle and is found at Devon Island. Finally, another famous picture, in which we can see the Rovers shadow and right next to it, the shadow of a man in a space suit apparently making adjustments with a hand tool device. Some UFO conspiracy theorists take this very photo and run amok with having a human colony on Mars, and the wild enthusiasts too caught up into believing every snippet of this which I do consider garbage, but the delusional state does get worse as a result. In summary, Devon Island has an identical landscape that is very eerily similar to NASA pictorials and images of purported of the Mars landscape and fittingly maintains a base there with Rovers fitted with cameras roaming around and people fully geared up in space suits. The NASA admissions about editing the photos before releasing them on their website and their NASA Channel livestream for public viewing, and the fact that NASA's web of lies is so very thick, and that, either by accident, or sometimes probably intentionally faking the landscape, and with some of the moon landings were filmed at secret locations, including Area 51, and the overwhelming consensus is that none of NASA's images are from Mars, not a single one !!!

NO PARALLAX- THE ISS-THE INTERNATIONAL SPACE STATION-

The definition of parallax-The difference in apparent direction of an object as seen from two different points. When you are fixing to land in a commercial airliner, you look out the window and the immediate horizon from the wing of your aircraft whether you are front back or rear peering out, you are surveying the speed of the horizon moving past you especially if one is determining the aircraft as it slowly descends down on the runway. But, a perceptive person may also look in front of the craft at the horizon for a definite distance of at least five miles and also notice that the distant horizon remains still and not moving at all. This is called the parallax of what is real and the fact that you traversed a flight that has been completed as the plane lands the craft and slows down to an eventual stop at the gate of disembarkation at the gate that it is scheduled to allow the travelers to leave and empty the cabin.

When viewing NASA's live-stream feed of the ISS, or as we say in the flat earth reality that our own eyes are seeing, the International Scam Station, we see the fake space station with the whole totality of the ball earth moving at one time. Their is absolutely no parallax to what you are viewing which debunks the reality of what you are seeing. To me, it's a bad CGI of what a second rate Hollywood movie would make. It is only if you are to use your keen sense of judgement to see right through this charade of fakery and trickery by the freemason frauds known as NASA. The 100 proofs that Earth is not a globe-William Carpenter,talked about Parrallax in what Proctor suggested in his works because he was known as the father of parallax and we will try to create the proper perspective of just exactly what he wrote about and what we can see and the ingenious of how his observations of this common sense science, and I do mean really scientifically provable experimentation that we seldom see in todays scientism of mostly made up science to fit the global-elites endless theoretical science to fit the needs to further create illusory made-up science to fit right into their agenda of what I like to call bullshit-science.

"With the LIRO experiment and with the conclusive theory of "gravitational waves" and the fact that most of the science of physics are suggesting that they know more about the universe than they know about the Earth". I have only one thing to say about this and it's similar to what these paid -off shills to spread the message that these globe ball believers have to continue to push.and this comes from the you-tube channel of Globe-busters. " Scientists have made-up more about the universe than what they know about the Earth. "-Jeranism

I saw recently that Elon Musk, who started the civilian SpaceX and has been funded over $4 billion from the U.S. Government to fool the fooled with the FakeSpaceX program, has been lauded a hero to the mainstream media for giving $ 10 million donation for the A.I. program through Google. It's astounding to me that these individuals are put on a pedestal to continue the fraudulent propaganda called space. I am under the impression that since the Super-computer at NASA's Goddard Center have more technology to be able to hoodwink us more because of the age of the "green screen" and the digital imagery with them is shared extensively with other branches of the Freemason "Hollywood elite directors" making the latest movies by making epic films like "Gravity", which left several surveys and comments throughout the social media thinking that the movie had actual real footage of space. I have seen at least three different sites stating as many as 45% of the viewers were convinced that some real space footage was used in this movie.

Jeranism is filled with data that seems unsurpassed by anyone and his social you-tube channel is the great insightful awakening that is happening world wide. The "Lie SS" has legally suffocated the taxpayers in our country the last eight years with over $150 billion which would solve the hunger problem throughout the world for the next 11 years, but instead, we are being taken in to something that is not based on reality. We are allowing reality to be faked and it is part of the in doctrinal poison that I was in as well for over 45 years of my life. We must take our reality back, because we are living in an imaginary bubble created by the controllers of money, commerce, transportation, selected free mason leaders, mainstream media outlets, everything around you has been created by these psychopaths who control everything literally because they are so very well-funded " By 2016,The Top 1% will be richer than the rest of the world combined."

> "The system of the Universe, as taught by Modern Astronomers being founded entirely on theory, for the truth of which they are unable to advance one single real proof, they have entrenched themselves in a conspiracy of silence, and decline to answer any objections which may be made to their hypotheses...Copernicus himself, who revived the theory of the heathen philosopher Pythagorus, and his great exponent Sir Isaac Newton, confessed that their system of a revolving Earth was only a possibility, and could not be proved by facts. It is only their followers who have decorated it with the name of an exact science,

yea according to them, the most exact of all sciences. Yet one astronomer Royal for England once said, speaking of the motion of the whole Solar System: The matter is left in a most delightful state of uncertainty, and I shall be very glad if any one can help me out of it.' What a very sad position for an 'exact science' to be in is this !"----David Wardlaw Scott-"Terra Firma"

AT WHAT COST ?

I find it utterly amazing, yet incomprehensible that most of these deceivers of man's truth in his own conscious to formulate breathing robots in their master-agenda that was engineered over the past 500 years rather brilliantly. These guys were all God believers, John Wesley Powell, Charles Darwin, Neil Armstrong, and the list is monumental as a fact. The thing that is totally reprehensible to me is the selling of your soul to become the ultimate 33rd Freemason and the rituals of pagan-Gods before entering a lifetime contract with the propagators of the master deception, and we all know its the Satanic order of the Freemason-Cabal that is literally applying cancer to destroy America and America's soul by attempting to rid and erase the cultural past and to speed up the global-takeover of the United States and I do believe that President Trump is resisting this more so than we are being told or led to believe. The New World Order (NWO) is coming and the speed of it arriving would definitely been on a speedier and scarier trek if Hillary Clinton had been elected, but I digress, the uncertainty created mostly by the largest hate group in the country, and that is the Democratic Party, in particular, the left-wing and the news media who refuse to acknowledge that the hate groups are coming from both sides, and no matter how you slice it, hate is hate and any form of hatred is appalling to any values, including American values but most of all, the value of the spirit of man and woman. We are in the times of hate, and revulsion,hating a President so much that a Senator from Missouri suggested on a twitter response that "She hopes that Trump gets assassinated ! She will soon be out of a job, at least in this country. The hate is everywhere and the robotic nature of human beings who are lack the ability to think for themselves and seemed to be almost mind-controlled. It's like the worker bee's who cannot do or say anything except for what their program calls for. This is indeed the age of deception. It's called the "New Age" for a reason, and none of it has to do with anything that's good, nor would it be true, and the headliner of decisions is what has brought humanity to this point of insanity and hateful caos essentially almost like a horror reality has settled in today. Some of the nihilistic nary of any good thought lies right in the lap of the electronic mainstream news media and also include all social media. The disconnection from you true reality has manifested a hurricane of hate that never passes over, it just sits there and grows larger and the sanity of any are manifesting hate as if they were all in some kind of trance or something resembling a zombie, with a singular goal to impose verbal and physical hate to create the pace of our own downfall, and both right groups like the KKK, and the Neo-Nazis, and it also applies to the ANTIFA-communist left wing radical group, and to BLM, because "ALL LIVES MATTER" and it should be called "ALM!!!" The parasitic agenda is literally destroying and allowing man to escape the love that only God can put back into man's heart. This nefariously evil deception comes from Satan and the army of Satan that rules over humanity is all to real and too symbolic to discern and to hopefully finds its sea-legs to prosper and grow, because all of humanity deserves to "love' and be loved", but to also give man back the truth of his own destiny and we are being denied this at every functional level possible by these leeches.These people keeping these-lies very close to the vest for riches and fame, but the inside of these God-fearing "gatekeepers" that are holding that lie close to your soul will eventually rot from the inside out instead of the other way-around! to disconnect mankind from the truth and also his potential both mentally and spiritually.

The very controversial subject of the cosmology reality that is accepted as fact has had many leaks over the years because the cost of being accepted into the 33rd Freemason-Illuminati highest ascension of there light is to delete your accountable resource of your outside spiritual dependable resonance, with a powerful lobby of self dependence on oneself to be of the greatest importance and to accept the narrative of what the highest order of Freemason dictates, and for all practical purposes, you are thus indoctrinated into the rituals of a Satanic order infested with a Satanic agenda. The Lie is a part of the secret societal order to maintain that secret at absolutely all costs. The Lies have been pretty much well kept for the majority of these astronauts representing the many space agencies who have been rigged into the official Freemasonry-Illuminati and the NWO have no boundaries between actual countries, but have divided the whole flat world into 33 sections, and you can count them on the United Nations Map. Interestingly enough, the UN Map doesn't show but six continents and the two golden fleeces at the bottom of the UN flag that wrap around the lower ridge acting as an ice wall holding the worlds oceans in and this crap is shoved in your face each and every day but, I digress, the acceptance of this poison doctrine has severed cognitive thought from most of us, and he who does not question and believe everything from this atheistic doctrine based solely on theory has put him or herself as a sacrificial lamb and will be almost hopeless to wake up from this phony reality that we were all infested with. Why not look no further than the purported first astronauts to walk on the Moon. Neil Armstrong and Edward "Buzz" Aldrin. When I was a space freak and looking back on it all, I was absolutely a space warrior at that time in my life, but as rubbish as it was, I am certain I was just like every one else who believes in space and of space travel, I was a "SPACE CADET." I lauded all of these 33rd Freemason Astronauts as the pinnacle of heroism. Neil Armstrong became a very mysterious person after his fake moon walk and disappeared without a trace and did not seek or want the limelight and not a single interview was ever registered live or taped of Armstrong reliving the exotic moment of being hailed as the first man who took a step on an alien surface called the Moon, or more appropriately, Lunar. The personal humiliation will almost half the time take it's toll on all of humanity. Neil Armstrong went into total reclusively from the public eye throughout his life and gave a cryptic message on the twenty-fifth anniversary of the Moon Landing Hoax in 1994. The clues of his guilt-ridden conscience was very evident in the short three minute speech that he delivered.

THE NEW AGE COSMOLOGICAL EXTRATERRESTIAL DECEPTION OF ALIENS AND THE REALITY OF WHO THEY REALLY ACTUALLY ARE, NAN- THE NEW AGE NEPHILIM-INTER-DIMENSIONAL FALLEN ANGELS;NOT SPACE ALIENS

I use to be in the same state of confusion that most everybody has been in at one time, and that is the re-branding of the watchers of the past, the fallen angels straight out of the book that man intentionally revoked from the Bible because it is very telling about the very rulers of our world then, as it is now, the children of the "FALLEN WATCHERS-ANGELS" who quickly gained control of our world because of their once heavenly powers who fell from the graces of God and used their inter-dimensional powers to subjugate the people of all humanity. The cloak of secrecy has been brilliantly defined in this gigantic and very mind-boggling deception that will undoubtedly define the end times of tribulation. We cannot have any kind of extraterrestrial's and their inter-dimensional magical shapeshifting UFO's without being on a global spinning ball earth. The disconnect from God has been almost completed with this super magical lie that almost rivals that of Flat Earth, and the ominous Darwin Theory, the Theory of Evolution, the equally repulsive "BIG BANG" Theory where a whole bunch of "nothing" turned into "everything", are the very same descendants of the angels of Satan. The same exact propagators of every war that has ever been fought on all soils of our plane called Earth, and are the rulers of our world because they literally created a system of the very virtues that they represent, power, greed, money, and conquest over all peoples of the world, in particular, people who love God like I do, and any Christians who rely not on any Religion, but all who have a personal relationship with Jesus Christ and try to emulate the way Jesus wanted man to live his life. The next great diabolical deception is coming to our false reality very soon, and it will involve the "Fake Alien Invasion", and it will be an easy excitable sale to the hypnotized sheep who refuse to go out of their cyanide box of a learned state of its own paradox. The mind is in prison and it is my subtle duty to break as many chains that I can, and my commitment to the truth of understanding the state of mind imprisonment is conclusively correct, and not one day will ever go by that I will ever subtract my own thoughts of the truth and you simply are freed from the cloudiness of truth like I was for over 50 years, but, I am going to always seek the truth` and everyday is so exhilarating to me and the comfort of knowing that I am finally free, and today have a sovereign mind, and will never allow the system of liars to ever corrupt and defame my mind from its true identity again!The Liars deceitfulness will end but not before the "fake invasion" that will be part of the fixture of our new reality pretty soon. I am going to revisit some of the phony misidentified cases of who these Aliens really are, and will refuse to document where they came from because the inconsistencies and lies coming out of the little green men's mouths are very consistent with Lucifer's army of fallen angels disguised as space aliens from other worlds. The Underground Nephilim Bases and Subterranean Tunnels of diagrammed below.

WHERE DO THE INTER-DIMENSIONAL NEPHILIM LIVE- THE UNDERGROUND SUBTERRANEAN TUNNEL SYSTEMS ALL OVER THE WORLD

AFGHANISTAN

"An ancient legend among the Hindus of India that chronicles a civilization of immense beauty beneath Central Asia. Several underground cities are said to be located north of the Himalaya mountains and have seen strange UFO Nephilim flying discs in Afghanistan and even more under the Hindu Kush. The subterranean Shangri-la is frequented by a race of golden inter-dimensional beings who don't communicate with the surface world. They travel thru an elaborate tunnel system that stretch like surface highways in many directions. Entrances into the tunnels are believed to be hidden in several of the ancient cities of the Orient. Some villagers have found tunnel entrances in Ellora, and the Ajanta caverns in the Chandore Mountain range of India."

BRAZIL

In Ponte Grosse that is in the state of Parana, is a tunnel that has been seen numerous times but city officials have been able to suffocate the opening and is off limits to the locals. The foliage of fruit orchards and very heavy brush has made access very hard as a result. Another entrance is near Rincon, which is also in the state of Parana. Also, in the state of Santa Calarina, Brazil, near the city of Johnville there is a mountain containing an entrance to the tunnel system. Some of the elderly villagers have told many stories about hearing singing coming from the underground caverns throughout Santa Catarina. Another area that is also in the state of Santa Catarina, just south of Gasper, have many unusual subterranean fruit orchards. Part of the beauty that outlines these areas are a vast network of honeycombed Atlantean tunnels that lead to subterranean cities.

ENGLAND

The Staffordshire, England entrance was discovered by a farmer who discovered a very large iron plate beneath the dirt. The hatch was large and oval shaped, with an iron ring mounted on it. This entrance led into the tunnel system and could explain why Staffordshire has been a hotbed for a lot of Nephilim UFO activity since the 1960's. The isolated field where the tunnel system is located seemed to be a perfect place because the valley is surrounded on all sides by heavy woods that keeps this spot an ideal area for a subterranean entrance. The laborer was digging a trench for purposes that were unknown. The folklore of this particular incident was reported in " A History of Staffordshire" by Dr. Plot, who wrote this book in the late 1700's. The difficulty in trying to find this entrance isn't impossible, but it still is undetermined exactly which valley the farmer was digging in.

CHINA

In Chinese folklore, we find lapses for hundreds of years, especially during the 20th century, which was the iron curtain years, and to try to verify some of what was told, is very difficult because, unlike today, it was hard to find out anything that can be verified as being confirmed to be truth. A book exists that's entitled "The Report Concerning the Cave Heavens and Lands of Happiness in Famous Mountains." written by Tu Kuang-t'ing, who lived from 850 to 933 AD. The book lists ten different 'cave heavens' that were supposed to exist beneath the mountains in China. Here is a report of an experience of a man who entered a passageway leading to one of these cave heavens.

After walking ten miles, he suddenly found himself in a beautiful land with a clear blue sky, shining pink clouds, fragrant flowers, densely growing willows, towers the color of cinnabar, pavillions of red jade, and far flung palaces.' He was met by a group of lovely, seductive women, who brought him to a house of jasper and played him beautiful music while he drank 'a ruby-red drink and a jade-colored juice.' Just as the felt the urge to let himself be seduced by these fallen female angels, a Godly ominous spirit allowed him to remember his family and he returned to the passageway. Led by a strange light that danced before him, he walked back to the cave to the outer world; but when he reached his home village, he didn't recognize anyone he saw, and when he arrived at his house, he met his own descendents of nine generations hence. They told him that one of their ancestors had disappeared into a cavern three hundred years before and had never been seen again.' Winston Churchill-England's Prime Minister during the second-World War and was one of the heroes that came back as a leader for the times and a beacon of leadership for Great Britain as well as the whole world, in a 1952 memorandum, inquired about a topic about UFO Flying saucers to Lord Cherwell, then Secretary of State for Air. Churchill's note said, "What does all of this stuff about flying saucers amount to? What can it mean ? What is the Truth ?-Winston Churchill The response to him indicated that everything could be explained with prosaic answers including meteorology, astronomy, misidentified conventional aircraft, delusions, and deliberate hoaxes.

A cave called, The Liyobaa Cave Entrance, as told by several Catholic locals, said that Catholic Priests had permanently sealed off this cave believing it to be an entrance to "Hell." The village of Liyobaa means " The Cavern of Death" and was located in the province of Zapoteca, somewhere near the ancient village of 'Mictian" translated to 'the village of the Underworld.' The Cavern of Death was actually located in the last chamber of an eight chamber Temple. This Temple had four rooms above the ground and four more important chambers built below the surface of the Earth. This building was so unique and was located in "Theozapotlan, and the tunnel entrance led one beneath a mountain."

EGYPT

Dr. Earlyne Chaney, in an article titled 'Odyssey Into Egypt' in her occult-oriented magazine Voice of Astara, in May, 1982, tells of a discovery she and researcher Bill Cox was shown in Egypt. There were two tunnels, neither of which had been fully explored. One was in the Temple of Edfu between Luxor and Cairo in the ruins of El Tuna Gabel; and the other near Zozer's Step Pyramid at Cairo near Memphis-Saqqarah, within the tomb of the Bull, called "Serapium"Soon after this discovery, the Egyptian government sealed both entrances because of fears of certain archaeologists who alleged that they "lead too deeply down into the depths of Earth," and because they found the earth to be "honeycombed with passages leading off into other depths," and the possibility of explorers dying and getting lost. If such labyrinths do exist, then it may explain a story that men dressed like "ancient Egyptians" have been seen in unexplored tunnels near Cairo, as well as possible confirmation of the story which appeared in Nevada Aerial Research's Leading Edge publication to the effect that the U.S. Government secretly maintains a huge base within a cavern of tremendous size. It is several miles in diameter beneath the desert sands of Egypt. Others suggest that it is a subterranean society referred to by certain people 'in the know' as the Phoenix Empire, or the 'Gizeh People.'

CANADA

The Nahanni Valley is a 250 square miles in the southern end of the Mackenzie Mountains, which is a mountain range in the Yukon and Northwest Territories, and hold over 55% of all the world's reserves of tungsten. Only two roads lead into the Mackenzie's and are both in the Yukon, and it makes for an ideal place for a Nephilim underground cave entrance. The Nahanni Valley have several Hot springs and sulfur geysers that keep this valley at least 30 degrees warmer than the surrounding areas that is typically very cold during most of the year.

UNITED STATES

Virginia

BELLS COVE-A ghost town, and a small near Buck Hill Caverns, deep within the Caverns some explorers discovered a seemingly bottomless shaft, or a pit. A number of them heard some loud but distinctive sounds of sobbing and crying of a woman wailing in pain.

Bluemont

In northern Virginia, is a virtual underground city loaded with 49 nukes from Washington, D.C., what is called the Continuity Of Government facility and is the hub of the FEMA-Subterranean network, and is called Mt. Weather. It is absolutely no surprise that all of these underground manned and interdimensional bases connect all of the major airports near and around the nation's capital. The infrastructure at Mt Weather is an amazing microwave communications system, and a small spring-fed lake, a pair of 250,000 gallon water tanks and several ponds stocked with fish. A sewage treatment plant capable of processing 40,000 gallons per day, a hospital, cafeterias, a diesel powered electrical plant, private housing and dormitories, and a massive super-computing facility which store information on millions of Americans, war game simulators, futuristic Electric cars. Insiders admit that an entire non-representative appointed "back up government" would live in residence within Mt. Weather, in case of all-out Nuclear War. There are several C.O.G. facilities within a 300 mile radius of Washington, D.C. with Mt. Weather being the central head coordinator of this "Federal Arc." There are at least 96 underground government C.O.G. that exist in Pennsylvania, Maryland, West Virginia, Virginia, and North Carolina.

Langley

At least 7 levels of underground facilities between the CIA Headquarters, some of which contain recovered interdimensional hardware from the Nephilim.

Washington, D.C.

THE House of the Temple-The Scottish Rite Masonic Headquarters- sits atop the pentagram-like street layout of Washington D.C. and reportedly the Capital connects to ancient glazed tunnels built into a very elaborate system believed constructed by the "Atlanteans" during antediluvian times, and is also connected to a massively huge cavern called "NOD" where the mistaken identity of these aliens called the "sirians" and the NSA, of other alien species, which in a phony reality of Hollywood and our present conditioned of a different state of who these entities are, and the powers-at-the controls state an agenda of aliens that are not extraterrestrial at all and the proof is exhibited every single day in our lives. The fear of alien worlds and creatures of galaxies afar, is science fiction in fact. It has no evidence to even ascertain where these 'Nephilim' come from and how many light years is 'Betelgeuse" may be, and it makes me laugh, because it is so very humorous and illogical for anything that these parasites suggest in psychotically based alcemistic science of the theory people! It is fodder for most who live half-conscious of the lies who believe and prey over each word as the truth as told and programmed to them by the hero's of in genial science, or the religion of Scientism, easily came from the doctrine of the Catholic church in Rome, and the astrological allusion taught when I was a child at St. Mary Magdalene, a catholic elementary school in Metairie, if you weren't going to be a "priest", then you are wanted to spread poison of even more monumental, and that is the study of science, especially theoretical science of the parasitic emplacement to line up the cattle and OBEY the nefarious lies from the cosmic belief in an heliocentric Lucifean 33rd Freemasonic-Illuminati doctrine that was planted to educate us into the delusion that we are in today. Science is always going to state the polar opposite of anything written in the holy Bible. Supposedly, the government and the Nephilim fake space warriors, have a collaborative Agenda for Global domination.

UTAH

In Salt Lake City, below Crossroads plaza behind the "Crossroads Cinemas Theators" are reported ancient tunnels that were discovered by construction workers and excavators, and some have been remade and refurbished. Not far from a reportedly Human-Nephilim collaboration in the area. Dangerous encounters with reptilian humanoid interdimensional futuristic technology with the federal agency involved, men seen in a 300 foot long chamber wearing suits and carrying Uzi machine guns, holographically concealed side passages, and metal gates to obstruct the scene itself. This gigantic underground city system connects the Zion Canyon, in Utah and extends into parts of Nevada, Arizona, and Colorado.

DUGWAY

Dugway is the new Area 51 according to most of the dogmatic narrative from several magazines and various science shows with their newest technology, mainly CGI pictures, but also the many years that the U.S. Government and other Western allies, but Russia interacts with these Nephilim entities to try to develop the technology that defies whatever magical portal that these inter-dimensionals enter into the realm of Earth and its living souls that is the bargaining chip to these beings who have the super natural ability to procreate into humanoid and of reptilian form in their engagement with our surface world with us. The underground facility connects Area 51 and Gloom Lake in Nevada, and is a sub global main post, with a couple people who have claimed to have seen automations and Nephilim reptilian humanoids opening under holographic human disguises, who have been seen to transform temporarily to their alien state, unfortunately, since 2008, Dugway has been condoned off in lieu of our government contaminating the area 51 site that's near Rachel, and the number of nuclear testing that was part of the Trinity project that was tested there from the Manhattan project during which led to the U.S. dropping the two bombs in Hiroshima and Nagasaki, Japan and shortly afterwards the Japanese surrounded to the allies to end the second World War. Dugway has become an additional and even more isolated area that will make the underground tunnel angels disguised as alien beings from Hollywood made up galaxies to further the evil New World Order Agenda that these beings are using as a bargaining chip in exchange for the technology that these Nephilim are promising to share with our military installations in exchange for their right to abduct some people so long as the abducting fallen ones return the abducted person after the experimentation and without anyone getting harmed, but this was never an agreement between anyone, and its my belief that this was another curve ball thrown by our esteemed misinformation units of our government, and therefore is just another well calculated lie, or sci-op. These entities feed off of the fear, and frenzy from the very minds of the traumatized victims, and that negative energy is powerful for their ability to maintain their shapes and forms.

ARIZONA

Grand Canyon- a large Cave near the confluence of the Colorado and Little Colorado rivers are a very popular Tribe of Native Americans called the Hopi Indians. In Hopi culture and old held legends, suggest that their ancestors once lived in the underground world with a friendly race of Nephilim that looked like "ant people" and not to be confused with the sinister attitudes of the Anakim "mantis" people who have been abducted, but some of their kind turned to "sorcery" and made an alliance with reptile-looking lizard Nephilim-Anakim-interdimensionals, and a serpent man known as the "two hearts." which dwells in even much deeper caverns below. The overwhelming fllod-gates of evil and violence forced the peaceful Hopi's back to the surface world. G.E. Kincaid claimed that he found an amazingly shocking "ancient cave" in which he reportedly discovered Oriental, Egyptian and Central American artifacts. The Smithsonian archeologists named S.A. Jordan and his team, explored the caves deeper into the cavern, and it had hundreds of rooms, enough to hold 50,000 people. The underground city is about 42 miles up river from El

Tovar Crystal Canyon and Crystal Creek and about 2,000 feet abover the river bed on the east wall. John Rhodes did over 3 years of intense research reportedly discovered the Grand Canyon city, which is now being used as a museum for elitist groups and has lower levels that are being used by "super secret black operatives" and stands guarded soldier dressed in white-jump suit and armed with an M16 rifle.

The Superstition Mountains- Several people claim to have had experiences with all types of Nephilim "alien" beings inhabiting the underground for which our creator set just for them, the demon enties of the fallen ones. This locale is under the Superstition Mts just east of Phoenix. Saome of the reports speak of humans and human type dwarfs inhabiting the caverns of this region, although in the 20th century there have been numerous reports of the Nephilim entities that look reptilian and the clone -like Grey Nephilim. Some reported also the white-greys who telepathically are telling our scientist that they are from Epsilon Bootes, their very appearance of half humanoid or a reptoid 'deros' in black hooded cloaks using abducted/programmed humans who mostly are spirit-less even before any abduction activity from these inter-dimensional s.Unfortunately the dark spirit world are only a probable event if the spirits are prompted as an invite. The cases of being abducted have made most of them fall into severe mind slavery or worse, reptilian Nephilim humanoids have integrated human DNA abducting and raping women and placing a chip inside some of them as a way of servitude to come in as they may since they were an inviting spirit that preys on the weak spirits of non-belief of being faith-less to a creator, or intelligent designer.The chip is a tracking device and is a way to manifest an inviting soul and spirit of emptiness to fill that void metaphysically and metaphorically. The events are too real, but so was the events that happened 4,500 years ago when these same entities of the devil entered the sky above Mt. Hermon as perfectly described from one of the Bibles omitted books called "ENOCH". Essentially, these inter-dimensinal beings have enough negative energy that is successfully transferred from the spirit world and feeds itself with negative spirit activity in the flesh world. The very strange and amazing impact on the very negative consciousness of an innocent is aplumb to taking a brick and killing the good spirits of your own make-up as being enlightened to good, because the lies and deceit, especially the convincing nature of believing in the fairy tale of astronomy and the cosmos of never ending space with no nature of God in most of these victims of demonic nephilim alien takeover, and it is mostly in a sub-conscious state whereas the probing and the documented trauma that this entails is very frightening to all who do not know better. These Alien beings are the army of Satan and has absolutely nothing to do with the pseudo Epsilon Bootes of being 65 million light years away and that they are 2,000 years ahead of us in technology and spew all of this garbage to keep you and I in space mode. which is so ludicrous and so doggone short sighted, but most of all, its based on psychotic theoretical pseudo science.

CALIFORNIA-

The Mt. Shasta and Death Valley area- The Paihute Indians of the Southwest USA claim that an Egyptian-type Nephilim race first colonized gigantic caverns within the Panamint Mts. thousands of years ago that some sources say could date back at least 2,500 B.C. which is approximately 600 years after the rise of the Egyptian Dynasty and its dominantly intellectual culture. Death Valley was part of an inland sea connected to the Pacific Ocean.

Deep Springs-A CIA paramilitary force agent called "yellow fruit, or yf", was based at the Nevada Test Site where he was working with "Blond Nordic humanoid Nephilim in a lie to form an effort to fight the Greys-other Nephilim fallen watchers, and Deep Springs is a nesting area for the Grey Nephilim aliens and the communist trancelike state homosexual human collaborators who are using the "National Resources Defense Council" as a front for their agency, sort of exactly how the Freemasons use the Catholic church as a front for their anti-humanism agenda! They allegedly are involved in an electromagnetic war with the so called "benevolent ones" who have made allies and have demon-possessed scientists at the Nevada Test Site, and serve as advisors to other intelligence agencies that are turning against the greys as a result of betrayals of joint-operation treaties, and the various inconsistencies of being truthful, on the contrary, the deception and telepathic powers of these inter-dimensional entities is only possible when an encounter is brings about fear, emptiness of self worth, depression, anger and hatred. These beasts prey on the wicked and the weak, and their spiritual endorphin's of evil only brings intensity through the ugly dispositions of thought propelled with terror and the fear that these abductions encompass, it's certainly not anything of the pure creator of our holy- God!

EDWARDS AIR FORCE BASE

The "haystack" bluff or butte near the launch area reportedly holds underground bases with several levels that is part of a gigantic elaborate system of connected tunnels that leads to other tunnels that connect military facilities and some airports all over the "purple plain majesty" of the U.S. Several witnesses have described alien-Nephilim activity and others have died under mysterious circumstances. It is also reported that a 50 mile underground tube shuttle (Disney shuttle comes to mind-above Earth's surface) that links Edwards AFB with the Tahachapi facility, and an ongoing excavation below the base down past 9,000 feet, with underground facilities being monitored by hovering remote-controlled basketball sized metallic spheres capable of electronically monitoring the brains, or the encephalographic waves (brain waves) of base workers and visitors to anticipate their intents. All of these actions has enabled these condemned inter-dimensions to coexist with our scientists and many military personnel who essentially are hoodwinked into a trance summons of working along-side of extra-terrestrial beings from other worlds in outer space! Our own government has engineered a lie on to itself. They have convinced these workers who are working along side these demonic entities disguised as aliens from Betelgeuse or other star-lights from the very ceiling of the translucent and plasma dome- firmament to formulate a concoction of an imaginary astrological based fairy tale to create a narrative of space-aliens that simply doesn't exist. These beings have been here since the Mt. Hermon event about 4,518 years ago. These super natural inter-dimension entities are programming the minds during every interaction with the flesh-beings who work with them. It has to be very excitable anxiety for the sheeple that work with this type of surroundings and coming out of one program to another to reassert mind-control over the breathing workers. I am guessing that some apprehensive fear is also involved to produce negative energy that actually feeds into the entities level of energy and thus is a regular conduit of energy for these beings. That is precisely why most all abductees have claimed that these beings are telepathic and it's because the vessel of energy starts with the MK ULTRA MIND-CONTROL PROGRAM that is already been in place since the late 60's.

Lancaster

A collaboration between Northrup, McDonnell-Douglas and Lockheed is developing and testing antigravity air and/or space craft in massive underground facilities, abductees report being taken to these elaborate multi-billion dollar underground complexes where they have seen military personnel with grey alien nephilm beings and even reptilian humanoids of nephilim. The glowing discs, triangles, boomerangs, elongated shapes, spheres and other types of craft have been seeing flying or hovering in the area. Sometimes you will also see jet-black unmarked helicopters.

California-Mt. Larson

a hiking party that discovered a glazed tunnel at about the 7,575 foot level of Mt. Larsen entered the cavern using a strange tunnel vehicle and were abducted by men dressed as surface people who were known as the "horlocks" which are humans under the total mind-control of some Nephilim force, only to be rescued by an underground resisence force. Some other incidents involved two documented abductions near Lassen by a human-reptilian collaboration.

The Mojave Desert in California

Many stories of underground natural shafts and pits and some shafts are definitely artificially built that leads tounderground caverns below Iron Canyon near the El Paso mountains which points northeast of Mojave. Reports of underground demon activity, automations, and electronic vortexes, all of which are certainly monitored by secret government agents that roam the area.

Georgia
Atlanta

A major underground Bavarian Illuminati facility constructed in connection with the similar base below the Denver International Airport. Like the DIA facility, the Atlanta facility is occupied by the cult of the serpent (human and alien Nephilim collaborators) and is intended to be used jointly with the Denver layout as duel U.S. Headquarters for the NWO-"New World Order" regional control, and for continued "Montauk" and also "Phoenix" project operations.)

IDAHO
Burley

George Haycock-Druggist and explorer, found a shaft that could be entered via a boulder strewn depression or sink six miles west of Burley, and about a mile from the main highway. It is presumed to be in the opposite direction from the river. Native American legends told of a demonic race that would emerge from a cave and capture women and children. Mr. Haycock reported psychic attacks and impressions of the Nephilim evil activities taking place underground. The shaft led to a long square-cut ancient horizontal crawl space tunnel with branch tunnels and a cave-in which he attempted to dig through, although he was experiencing unusual "resistance" from very powerful force that he didn't see but felt it, and it was immensely powerful. He later wrote to friends that someone was trying to blast the shaft closed using dynamite and also reported a "death threat" that he had received in the mail telling him to cease and desist his explorations. Shortly after this, Mr. Haycock was found strangled to death in his home.

Louisiana

Fort Polk- Seems that these Nephilim and secretive forces are mostly tied into the governments and the military establishment. Reports of over

19,000 war-ready United Nations Organization troops, French, Pakistani, and Russian, along with massive underground facilities for storage of military and other supplies.

Maryland

Fort Meade- "Cavernous subterranean expanses" existing beneath the National Security Agency's headquarters, filled with over 10 acres of the most sophisticated supercomputers money can buy, which monitor global telephone, telex, telegraph, fax, radio, TV, microwave, internet, and all other forms of communication.

NEW MEXICO
DULCE

The Jicarilla Apache Indians believed that their ancestors emerged from the caverns in ancient times. When they emerged they were plagued by monsters., who called themselves the saurian, another lie by the demon entities waiting for the rise and return of Lucifer because they are the army of Lucifer. The Jicarilla fought many battles with these soul-less malcontents and were able to drive the Nephilim-saurian back underground, but several hundred deaths occurred in a prolonged war with these beings. In an ironic twist, The Dulce underground network is under the mind control from these same monsters or reptilians-Nephilim.

In the town of Dulce, New Mexico,is the Jicarilla Apache Indian Reservation which is one block from the Pan Am building where the old high school use to be located. It's now used as an engineering facility by Maken & Hanger and inside this facility is an elevator that leads to the infamous Level 1 of the massive underground facility beneath the Dulce area which is also known as "Ultra" or "Section D" . It runs under main street at a depth of 210 feet and is guarded by Proforce Security whereas deeper and more secure levels under the archuletta mesa to the north contain automatic devices designed to kill any intruders on sight. Dulce is easily the most massive and the most strategic of all of the "hubs" of the joint military-industrial/Nephilim "alien" imperial collaboration in the

U.S., with numerous tube-tunnels radiating to all parts of the continent and beyond.

The Dulce underground Military Base/Laboratory connects with the underground network of tunnels which honeycombs our whole world, and the lower levels of this base are allegedly under the control of Inner Earth Nephilim beings that have control and have convinced their co-workers who are essentially captured permanently by total mind control that has been shared with our government programs in order to infiltrate the ability to enact the MK-ULTRA mind control and have implementing at least 90% of this programming through the telepathic communications that our underground workers engage daily with what they think are aliens from outer space with technology that these Nephilim -alien beings are suppose to be sharing with us pathetically backwards human's which the narrative is that these beings are 1,000 years ahead of planet earth in super-craft called flying saucers, and the ability to communicate telepathy and to be able to read minds and are just so delightful, especially the benevolent races and that we should fear the malevolent beings from the planet ZORKO from a galaxy not yet named, but NASA will soon paint it for you ! In a more serious tone, I do know that magical aircraft can appear in different forms and even shape-shift while these fallen angels are in flight, but coming from outer space when space simply doesn't exist, only asserts the fact that these beings come from portals in the sky and aren't coming from 165 million light years away through wormholes. The fact that a wormhole is simply a portal that these dead fallen angels who have super natural powers because they are awaiting for the anti-Christ to be unleased on humanity, because they are going to be the main army of Lucifer. The base is also connected to the Los Alamos research facilities and its also through a sophisticated tunnel system that interconnects North America without even going up to the surface world of Earth. The shuttle system is a hypersonic train that is absolutely astounding and the general location among some locals is pretty well known .

Beginning in late February 1947, a road was being built near the nefarious government/Nephilim Base, under the cover of a lumber company. According to a couple local sources no lumber was ever hauled, and the road was later destroyed and dismantled. Navajo Dam is the Dulce Base's main power source. Most of the lakes in and around Dulce were man-made "government grants" for the Indian reservation. A machine that is called "The Subterrene"-was developed at Los Alamos and called the Los Alamos nuclear-powered tunnel machine, to continue new construction projects called Deep Underground Military Bases (DUMB), this machine is so fascinating that it burrows through the rock, deep underground, by heating whatever stone that it encounters into "molton rock", which cools after the Subterrene has moved on and the end result is a smooth and glazing lining.

Bechtel is a super international corporate octopus, which was founded in 1898. Some refer to the firm to really being a 'Shadow Government's" working arm of the CIA. It is the largest Construction and Engineering outfit in the U.S. and also the whole world. Many of the former Bechtel officers usually move on to some of the most important positions in the U.S. Government There are over 100 secret exits near and around Dulce. Many are located around Archuleta Mesa and Dulce Lake. Deep sections of the underground complex connect into natural cavern systems. In the realm of underground that is usually magnetically controlled electrical system that power elevators, lights, and all of the doors of the Dulce Base. Another happenstance that continues to this day is the large amount of animal mutilations, especially cattle. Some

of the cattle would lie lifeless with all of its blood drained in its carcass and missing its tongue, brain, and heart. The mysterious belief of superior technologically tools that are being used to drain cattle and taken out the vital organs without any secondary trauma from the cattle or the unbelievable precision of some kind of lazor cuts without leaving any trauma to the dead carcass,only amazement from the narrative of experts who have done examinations and autopsies of the lost cattle. The researchers at Dulce Base have also abducted several people from the local civilian population and implanted devices of various types in their heads and bodies. It wasn't until Livermore Berkeley Labs began producing blood for the Dulce Base in 1986, is approximately when the human and animal abductions slowed considerably. This group, along with the DELTA (National Recon Group) are responsible of all Nephilim "alien" connected projects. The DELTA symbol is a black triangle on a red background, and the irony is that the Dulce Base symbol is a Delta (triangle) with the Greek letter "Tau" within the triangle that is inverted upside down. Many victims have clearly remembered seeing this very symbol during the time of any hazy recollection from their frightening experiences.

NEW YORK

Manhattan- A large triangular system of tunnels utilized by a Freemason Lodge runs deep below the surface of Manhattan. It's reported that another opening to these underground tunnels to interact with these "beings from Hades" is below the Empire State Building. Not too far away, a worker named Con Edison, while drilling a test -hole in the north end of East River Park, broke through an open space about 200 feet below that led to a cavernous opening and was eventually closed back up and re-enforced from any outside curiosity seekers.

We will also find some strange stories of several sub-basement levels descending below the Empire State Building, and rumors are that these sub-basements are connecting to ancient tunnels and sub-basement levels that is under the control of the federal government. Following the Twin Towers bombing it was reported that 6 sub-levels controlled by the secret service were damaged. Also, in a similar fashion, local reporters in the early hours of the Oklahoma City bombing reported that underground tunnels containing all kinds of military hardware had been blasted open, and also rumors of an 18 leveled underground facility below the Murrah Federal Building, with the 5 upper levels being used for parking. These aspects of the bombings were, to no ones surprise, ever reported by the national media monopolies. Another story in Manhattan concerns The Episcopal Church of St. John the Divine, where it's reportedly an entrance to another underground cavern used by a nefarious masonic lodge, below this ground floor lies an actual city left over from ancient sub-inhabitants of the eastern seaboard during antediluvian/Atlantean times, and protected by space-warping Nephilim technology.

MONTAUK

An underground with at least 9 levels can be found deep below Camp Hero at Montauk Point in Long Island. This area is occupied by higher inter-dimensional beings who are oftentimes ruthless when they are encountering any U.S. Government officials. They claim that they are from Alpha Draco, and have possessed for 4,000 years just how easy it is to convince even mid-level operatives from our government that they are sharing technology and are spacemen from another faraway planet. The blanket of deception is very sophisticated and isn't lost on these Nephilim/Anakim creatures of the devil, and also, the Greys from their imaginary planet of their origin, Rigel Orion. Accompanying and interaction with these beings are handled by Black Ops agents, and German intelligence, which actually are the "Thule Society" agents, who are interconnected with the elitists of all elites, who only number about eighty-five, who actually knows what the whole entire "proverbial pie of deception", and the Agenda that has been in place for almost 2,500 years. The Montauk mind control and space-time manipulation projects were based there and at least 25 other bases around North America and reportedly involved over 25,000 "abductees" who have been programmed to serve them in their MK-Ultra program to serve them as 'sleeper' agents for the New World Order (NWO).

To summarize this in my own thoughts because for the first 54 years of my life, the thought of space exploration and the fact that growing up in the 60's allowed me to get saturated with all of the Gemini and the Apollo moon projects. I didn't realize that then, as I do now, that the race to space against the Soviet Union was an orchestration that closely resembled another, which was called the "cold-war". When the true events of how the unity became more than platonic with the Russians, is when Real Admiral Byrd found the true identity of who we really are, and, it scared the shit out of these Masonic ally attached world governments, the eastern and western blocs were put collectively on a secretive world stage to annex Antarctica and the time to do it was in its immediacy is not to be taken lightly, and the money barters from the elite were going to make sure that Antarctica was not to be owned, explored, not to have any man or company to drill the vast resources, and to also protect anyone from anywhere, from trying to navigate any flight or ship to this sacred land of Antarctica. The hollywoodesque sci-ops concerning Antarctica and the fact that a Nephilim UFO crashed in Roswell, New Mexico absolutely provided the story-line to give these "fallen angels" of its origins, the 200 fallen watchers that has been dissimilated from truth in the only truthful book that exists, and to actually rebrand these demonic entities whose supernatural characteristics and coming from portals from the 4th dimension and appearing to feed off of negative energy and are interdimensional in nature, not extraterrestrial, that is a fabrication built and developed by the same elites who have literally castigated lies by the thousands and especially since the birth of NASA itself. A quote from Samuel Clemons below.

"History is strewn Thick with evidence that a TRUTH is easy to "kill," But a LIE, well told is IMMORTAL !"-Mark Twain

I am no longer enslaved by man's vision of what reality he placed all of us in through alcemistic theoretical physics that eventually placed our very conscience up to the heavens, but not for our God, but aliens and infinite space and believing that the whole bunch of absolutely nothing, exploded and created everything is such a barbaric idiotic thing that isn't even plausibly possible. How can a "WHOLE BUNCH OF NOTHING" turn into a smashing "BIG BANG" and then the world of nothing created everything? I would think that someone has to be delusion of thought, or they are on some kind of LSD trip! So many people are now asking questions that never use to . This UFO-ET phenomena is Darwinism on steroids, the similarity between the two is evidence of even furthering your true conscious spirit that continues being filled with the lies of man, in particular, the elitist of the elites. This cobweb of deceit eventually created the disconnection of the spiritual realm that exists in every man's soul. The creator came back in the flesh, and Jesus Christ is not a fable, thus, it is the living God that has been suffering from the passiveness of man to be able to reconcile and the only way is for man to acquire a diligent rescue of his own mind that was taught a fairy tale of lies. Jesus is, the savior of man who provides his spirit that never walks away from us, but we sure dispel or forget him and thus, we walk away from the holy spirit. I, for one, will be given the courage to standup to this deceptive lie, and help my brethren to get out of the prison of our indoctrinated toxic lies that is our belief system. My favorite inspiration to my writing over the last year was the Antarctic Warrior- who had a you tube channel that featured "learning to fly" from the great rock band, Pink Floyd. It is unfortunate that his channel, or at least that video was taken down. The words to this song and the awesome footage of flat earth from a weather balloon of over 133 kilometers is breathtaking to see our round beautiful flat as a pancake earth, and at the end shows footage from a Hennessy Cognac commercial

Subject: Part 3 Summary: The Flat Earth Truth

SUMMARY TO PART 3: THE FLAT EARTH TRUTH

As of June of 2017, it is estimated that well over three million people around the plane are FET (Flat-Earth Truth), and the number is growing at an enormous rate because of the accessibility of the super-information age and when you discover that wool was placed on you to blindfold to believe in theoretical physics of what we know very little of and allowed this artificial education to indoctrinate you from reality. We thus have allowed science of what we know nothing about to rampage over the science that we know very much about, such as electromagnetism. Most truth seekers will always try to share the truth with everybody that is in his or her little family and friend circle. I always share my thoughts nowadays only when I am being questioned. I have many friends who constantly send me text messages or workers who are very inquisitive about the flat Earth. It actually has gotten to a comfortable point with my own circle of family and friends because many times, my coworkers tried to debunk me with their global theory and belief system and have been unable to do so, and it is not a surprise, yet, one must be careful to engage with a globe-head carefully with lots of intuitive thought. You must be compassionate to their feelings and humbly bring out the facts of the flat-Earth. It is particularly hard to try to wrap your mind that has convinced your every thought of the

Star Wars allusion that is very hard to escape and that's mostly by choice and pride.

David Wilcock of *Cosmic Disclosure* and *Ancient Aliens* lore, a new-age galactic space cadet and Alien disclosure and the fictional colony on Mars, does remind me of the old *The Honeymooners* show with Jackie Gleason, Audrey Meadows, and Art Carney, one of the all-time great comedy shows. Wilcock reminds me of one of the shows when he (Wilcock) is Art Carney's character named Ed Norton, Ralph's next-door neighbor whose TV set was broken. He was watching TV at Ralphs with a space-helmet on his head, and the show was called *Captain Video*. When Ralph comes out of his bedroom and encountering his uninvited neighbor sitting on his chair, watching his space cadet show, it was so startling, and the comedic skit was really a golden time of television. It's so unfortunate that the original Hollywood actors are mostly gone, but the late '50s actually was during the time that the official space hoax was being created. Here are David Wilcock's own words that he said on making a response YouTube video with regard to flat Earth.

> This Flat Earth thing is a fucking joke! If you walk from one side of the earth, to the other side in a sail-boat, you're going to wind up exactly where you started. You're not going to fall off the edge.

I'm sorry, but this whole thing looks like as a deliberate con from the NSA groups that lure in certain individuals in our community, who have been lured in with a narrative that makes them look troglodyte and ridiculous to everybody else in our community who are maybe looking for the truth. He then talked about Aristotle and his theory using two sticks, one 160 mi. apart to prove that the earth was moving around the Sun. Many Greek philosophers suggested that it was nonsense and counterintuitive to our very own senses. He finished his tirade of nonsense, especially losing over 100,000 subscribers from his UFO-space colony make-believe bullshit by adding you can use a sextant to determine angular distances to prove his point, which, again, is not very bright because, a sextant is good for navigation but proves very inaccurate on distances of latitude but proves halfway reliable on angular distances. He then added, "None of this stuff is true, you can see the Moon rotating his gesturing spin of his hand to illustrate his asinine point. We do not see any rotation of the Moon at all, and never will we! This paid off shill along with his other fake space warrior named Steven Greer are absolute delusional airheads! He ended his idiotic rant by suggesting that the NSA Government submersible's formed Earth groups and that their followers were also government agents from the NSA. My one question is this: how did this moron ever get on any TV or any programming? He then retorted angrily: "This flat earth is a joke and is just fucking nonsense! To me he seems almost unglued, and looking like he just woke up, or hungover!" (rants from fake space mouseketeer, David Wilcock)

"You're an idiot. You're a moron. You're retarded."

are not proofs of the globe-

WayKiWayKi, flat Earth writer, commenting on David Wilcock's tirade

This is coming from a guy who suggested that the Earth is a "rehab colony." This paid-off shill of a malcontent mostly offers zero science, only a filthy mouth laced with vulgarity and insults, even worse than Tyson's dropping of a microphone to prove gravity. The only thing worse than this scientist wannabe space cadet in the belief that he has "He Believes In Places He's Never Been to, and Rejects Where He Is (WayKiWayKi, Flat Earth Science.Org).

The Earth looks flat, tests flat, and the Earth measures *flat*, but I want it to be a ball. From Facebook photos of Flat Earth Revolution and Flat Earth Research websites that I have been able to provide snippets of truth and share with the ever-growing numbers of the brethren of sleeping sheep who are waking up to numbers that were over twenty-seven thousand per month, and that number has more than doubled in the summer of 2017.

Many of the critics have come out of the woodwork and claimed the flat Earth is a conspiracy, and they cannot be more wrongfully inaccurate in this assumption. For two thousand five hundred years, every culture that ever existed—the ancient Egyptians, the Norse, the Hindu, the Mayan, the Incas, the Navajo, the Hebrew, the whole Western world—were convinced the world was flat. In 1773, Captain James Cook sailed not once but thrice to the Antarctic Circle. He sailed for three years and three days for a total of over sixty thousand miles to the ice wall and was looking for a break in the wall itself, a mere break or a passageway that would enable him and his crew to try to discover any interior of the imposing wall of ice that also came with boisterous winds and seas of anger that can and will topple any vessel. The harsh hurricane conditions make any trip to the most challenging explorers, and Captain Cook was every bit the eighteenth century's Adm. Richard Byrd during his four polar expeditions that he made, especially the last two. Operation High Jump, in my belief, was when Byrd and his military support, including an arsenal of navy fighters and jets, took off for Antarctica's harsh interior and flew over 200 miles but discovered the plasma dome firmament. I could be subjective about this when I believe that a couple of our military jets crashed into the plasma dome firmament, and it is the same as crashing your plane into the oceans of our world. The sci-op part of "operation high jump" was that Nazi UFOs were responsible for the crashed planes, but the CIA-NSA and all the highest levels of the U.S. government are as shrewd as the Mossad. Of course, a lot of what was learned came from Scotland Yard.

Another false narrative about Byrd's polar exploits came from the Freemason Byrd himself, in a TV interview from the Wittnaur watch science show shortly after his third trip looking for the south pole, where Byrd stated that, right past little America lies an unexplored land the mass of the United States. This is but a trick and a "lie" that was a sci-op to confuse and befuddle most of the world to what their true intention actually was, and that is to hide the existence of God. They already knew that. Only they had the capabilities and vast resources to do just that. Not too many people have ever been to Antarctica, and we know very little of just exactly what Byrd found, but the false narratives and the "hollow Earth" science fiction stories were all part of these elitist psychotic agenda. They are the money hoarders of the world, and the agenda is very clear-cut. It is the opposite of what we deem as spiritually good and praising Jehovah, who was the only professor of truth. They literally turned the Bible into an antiquated fairy tale, but they even disconnected your subconscious state to gradually create a permanent disconnection of belief in the Bible or of any existence of God or the perfect word of his words and the only truth to humanity and helped create a whole lot of confused Atheists. That is exactly how the master deception works. The author of this master deception is Lucifer, and his minions such as NASA and the other Freemason dominated space agencies of our world.

> It's better to get hurt by the truth than get comforted with a lie! When a well-packaged "web of lies" been sold gradually to the masses over generations, the truth will seem utterly preposterous and its speaker a raving lunatic.
> —Dresden James

Local spot over Lake Pontchartrain, New Orleans.

The fact is that Al Burundi ancient flat map that dated to around the year 1,000. Furthermore, the most accurate map in modern history and has been deemed factually and scientifically correct is the Gleason map from 1892. The very first globe was not even introduced until 1492, and the flat-Earth map was just balled up unto a globe and so many inconsistencies still exist to this very day, like the loss of over 440 miles that has been never explained by any modern astronomers, only ignored! The maps back then reflect so much more accuracy, and all these cultures and peoples from around Earth's plane knew the world was flat even though they were thousands of miles apart. The Earth is motionless just as your own senses suggest to you and has been proven through many scientific experiments that are actually being suppressed as much as possible. Albert A. Michelson and Ed Morley were two of the very best well known and respected scientists of their time and, in 1887, with the most scientifically accurate experiment ever conducted, reached a consensus after using the speed of light to determine how fast our Earth is rotating on its axis. I recently purchased a book about William Carpenter, who wrote *One Hundred Proofs That Earth Is Not a Globe*. The outrageous price is set so you are dissuaded to paying $200 for a book that is only forty pages long, but the effectiveness of common sense reason and scientifically proven through experiments and not how it's done today through delusional equations through math to formulate fact. That is absolutely ludicrous, and technological age of deception has picked up tremendously. The fooled will be fooled even more so but also kept in their own state of stasis to something that's not a reality.

In a real downer to my very core of who I am, I am very aggravated that our young children are being poisoned into the same delusional reality. I can remember when I was at the University of Southern Mississippi in Hattiesburg, where ten years later, Brett Favre sprouted wings for himself and flew into the hall of fame after a fabulous career with the Green Bay Packers. We had one of the exchange students suggesting flat Earth to several of us at the student union. It was my sophomore year in November 1980 when a student got up and was very knowledgeable in his arguments with the flat-Earth model but lost a lot of credibility because in my eyes, as so with the others, he was invoking the Bible even though he was correct in all of his assertions on the flat-Earth truth. He kind of lost all of us and left us thinking that he was just a Bible fanatic. I will never forget the surprise encounter, but he did get me thinking about the same things that he had said. Now looking back on it, he was rightfully correct the whole time, but his missing ingredient was that he didn't have any stage presence and also gave the impression that he was a so-called holy roller! I came across something on Facebook that J. Edgar Hoover said, and it is such a resonating quote about flat-Earth truth.

> The individual is handicapped by coming face to face with a conspiracy so monstrous he can not believe it exists.

Another capsule to a summary should only be fitting as explained by none other than Eric Dubay. Eric is one of the heroes to the flat-Earth truth but has been involved with probably not staying on point and finding the deep rabbit hole that he continues to dig down to the levels of information of newer conspiracies involving Adolf Hitler and the things that just make the rest of the flat-Earth truth seekers lose a little credibility. All in all, he is one that is a tremendous shining armor to his considerable knowledge of this awakening, and his explanation says it all. He puts a pretty good spin—no pun intended—on the reality that we all face and the greatest lie in the history of mankind has been successfully executed. To that end, man is waking up so fast to the lies of it all, and the steam will continue its course to wake up the masses. I pray to God right now that all the heroes to our awakening to the truth, Matthew "Math Powerland." Boylan, who, along with Dubay, were the frontrunners to the new awakening, and I like them both. I could care less who is a shill and who isn't. I care about collaboration on every facet of resonating the truth! I believe the duty for all of us is not who is on top of the mountain of truth and knowledge, and it's neither of you two gentlemen. It's the God and creator who created us equally to his word and his truth! Man, we all have to just simmer down and find our common ground that will unite us despite some differences. The prideful nature of man is always leaking out of our endorphins, and we should acknowledge that we all have a particular duty to be able to group together as the official awakened sheep to push and tug and provide the truth to the sleeping sheep that our very senses of righteousness convey to us. Sometimes a prayerful moment is a recommended remedy to know that we are all the truth of our very destiny.

Below is Eric Dubay's take on our erroneously disjointed education that was handed down to all academia from the globe-elite Freemason liars. Mr. Dubay is absolute in his brilliant oratory of who did this to us and why.

> Wolves in sheep's clothing have pulled the wool over our eyes. For almost 500 years, the masses have been thoroughly deceived by a cosmic fairy tale of astronomical proportions. We have been taught a falsehood so gigantic and diabolical that it has blinded us from our own experience and common sense, from seeing the world and the universe as they truly are. Through pseudo-science books and programs, mass media and public education, universities, and government propaganda, the world has been systematically brain-washed slowly indoctrinated over centuries into the unquestioning belief of the "Greatest Lie of All-Time." A multi-generational conspiracy has succeeded in the minds of the masses, to pick up the fixed Earth, shape it into a ball, spin it in circles, and throw it around the Sun! The greatest cover-up of all time, NASA and Freemasonry's biggest secret, is that we are living on a plane, not a planet, that Earth is the flat, stationary centre of the universe.

I responded on a Facebook video that had my friend asking questions on Toilets flushing counterclockwise as our hurricanes do in the Northern Hemisphere and was wandering if the toilets in the Southern Hemisphere were flushing clockwise as the typhoons do? One of my old classmates in high school responded to my inquest, stating, "Why would it?" When he suggested "It's because of the spin of the Earth." Well, you know the door of truth opened up to rid this old buddy from the very room of deception, and here was my second reply!

> If you believe the Earth is spinning 1,100 miles per hour near the equator yet your own senses are telling you that you haven't moved an inch since you were born. The suppression of truth, especially since the world's two best scientists in 1887, Michelson-Morley experiment, along with several other experiments that levied out the truth, in

133,000 feet above, flat horizon, no fish eye lens.

lieu of Albert Einstein who has zero inventions to his credit, and improperly discarded ether which we know is part of space, and to make his theory of relativity work, he cheated, and which consists of proof by equational assumption of an imaginary fairy tale that we are a spinning ball earth which is contrary to our common senses, but we know NOT better. Yet "no man" has ever seen ONE AUTHENTIC REAL PICTURE of what this place actually looks like, NASA-HOLLYWOOD-DISNEY is one in the same, it's all CGI, cartoons and paintings to create NASA's yearly Earth images from space. It's as though you must unlearn what you learned! When we allowed Psychics and theoretical conclusions to provide facts, but were only fanciful lies and to allow the unknown of Psychics to trump over what we do know that we can assert as factual, that is, the reality of electromagnetism and the what we do know and have proven tom be real, has been superseded by a madness that have put most of the masses in a star wars allusion that doesn't even exist! Space is as fake as a $3 bill. Your satellite photos come from high altitude military craft, as well as some commercial airlines that come out the body and the belly of the plane to provide non-stop imagery from fitted cameras. Then the digital imagery provides the terrain from not satellites in fake space, but from planes. NASA and the 33rd Freemason Frauds is a gigantic tax scam just as the fake curvature of Earth is done by a "fish eyed lens Go Pro convex /concave lens camera. The three main qualifications of being an astro-NOT are

1. You have to have a bachelor's degree in Scuba Diving
2. A 33rd degree Freemason Illuminati piece of rich shit!
3. The great ability to LIE!

That's One Step for Man, ONE GIANT LIE for Mankind! (Mark S Hollander)

My second inquest to further my point was this: "They call it the American Dream because you would have to be asleep to believe it!" I know that I'm certain that there is no such possibility for the convexity of water in our oceans, and the navigational instrument obviously would be a small globe model with their proposed meridians being a semicircle. However, the mariners' instruments, especially the compass, points north and south at the same time, which proves that the meridian follows a path of a straight line. To be more specific, a horizontal flat line. When a ship is navigating the oceans and seas of the world, the captain's math equations is perfectly symmetrical with a flat Earth map for coordination on the assumption of 100 percent complete accuracy to complete his voyage or destination. If the Earth was a sphere, such an assumption would lead to glaring inaccuracies and the necessity for using spherical trigonometry. Sailing has been fine in both theory and practice for thousands of years. By using just regular plane, trigonometry thus proves more accuracy. In practice, scarcely any other rules are used except the ones derived from plane trigonometry when sailing. The conclusion despite some objections to plane sailing but both latitude and longitude are found most often and most accurate by assuming the Earth to be flat, much more accurately than being spherical. One of the most fascinating proofs of our world not being a globe—and it never was a globe—is the esteemed Swiss-born Belgian physicist Auguste Piccard, who is well noted for his exploration of both the lower troposphere and the upper stratosphere and actually was closer to being a real astronaut than what we are seeing today. Piccard actually ascended just over 9 miles above the horizon or actually 51,200 feet on the morning of May 27, 1931, and was conducting real scientific experiments. He was a tremendous asset in the study of physics and chemistry and was interested in balloon ascents as a means of making experiments. He partook in many studies and became a genius on the study of cosmic rays in the lower to upper atmosphere. In 1922, and soon afterward, he was offered a chair that was created just for him at the University of Brussels, and he accepted the post. He was the first scientist to recognize how fatal the stratosphere would be especially with the low pressure, and it is when you ascend right above the 7 1/2-mile barrier just above the troposphere. His balloon was airtight inside the cabin and was equipped with pressurized air, and it revolutionized the possibility that was soon realized for airplanes. I do know that Auguste would have been very proud of his grandson Bertrand Piccard, who was successful on the first nonstop round-the-world balloon flight in 1999. The Hennessy commercial that ran in the spring of 2016 featured real footage of Piccard's ascent over the clouds moving toward the stratosphere in what was visually seen as a flat as a pancake Earth, but the globe-elite company went a few steps further on what the viewer was watching, and we saw the globe-ball of Piccard's FNRS-designed cabin with the balloon disintegrating in the upper atmosphere and the ball-cabin rapidly being heated through the Van Allen Radiation belts but eventually moving past them and going through the big dome in the sky called the firmament. Obviously, a pressurized cabin would be unable to go through the highly charged particles that is approximately 99 miles to over 1,200 miles in the thermosphere where temperatures can get up to 4,500 degrees, which would have cooked Mr. Piccard. The true narrative on the commercial is, in my opinion, the Freemason elites "mocking humanity" on their mostly well-kept secrets that will blow humanity away from a metaphysical as well as well as redefining just how unique we are that encompasses our self-worth and knowing that a presence so superior in thought and uniqueness that falls on a renewed belief in God. An amazing article in *Popular Science* in the August 1931 edition: the very words will provide you with hopefully a transformation to really look into flat Earth and then try to debunk it by being an advocate of truth to be able to look at all avenues and being an accomplished deep critical and understand the whole canvas of thought, not just the part that you are comfortable with. Auguste Piccard is so highly regarded that the *Star Trek* movies, when it went to the big screen, named the captain and commander after this very astute and one of the most brilliant chemist-physicists that the world has ever known.

It seemed a flat disk with upturned edges.

I will try to characterize some very interesting conversations that have had all over the world using social media. Some were pretty contentious, but from the very beginning of breaking my own mind away from imprisonment, the golden texts and the great people throughout the world have been on their own fishing expedition for the ultimate truth, and my own guidelines have been an evolutionary process that literally will never end. I recently scored a 164 IQ test and aced it so impeccably that I thought of it as a scam of someone just trying to sell me something, but I digress. My revelation of the flat-Earth truth is a curse for those remain asleep but is the brightest of all blessings because you start to digest truth in every single endeavor in your life, including your job. I am going to share a few texts from yours truly, which I have come to think as hilariously easy to take on ex-NASA and Martin Marietta employees, including a satellite engineer and one rocket aerodynamic engineer who worked in New Orleans East on the "booster rockets" for the Space Shuttle program. I found this exercise almost too easy to absolutely annihilate their imaginative absurdity of their Star Wars belief in space.

TEXT

We all need to take our minds back, especially the crap of theoretical science of psychics (only 4 percent of what is known) was somehow able to trump over what we do know (what Tesla said), which is electromagnetism the governments of the world has intentionally failed us in education. I do mean indoctrination with BS and has created humans into a "race of idiots!" The very word "government," "govern," is an old Greek word meaning "to guide." "Ment" is an old Latin term meaning "mind." To guide your mind or "mind control." I often refer to people who cannot or will not get out of this mind imprisonment as sleeping sheep. The great awakening worldwide is happening, as people are waking up by the hundreds and thousands each day. Waking up to the lies of government-NWO-parasitic thirty-third Freemason-Illuminati just like the frauds from all the world's space agencies, all Luciferian thirty-third Freemasons and all but four countries—Libya, Syria, Cuba, and North Korea. Of course, we recently bombed Syria and probably going to war with North Korea. This is so the Jewish banking families/Vatican can introduce the WCB (World Central Bank) to control and force a "one-world currency. (Mark Steven Hollander)

TEXT

I don't go to church anymore because religion is part of the big problem, but whatever your Creator said in the only book that man cannot and never will be able to "debunk" is the Bible. In fact, all the prophesies are coming true right before our very eyes! Whatever God says, the world's biggest religion called science or "scientism" says the exact opposite. Who do you believe? Man are known liars. They have made the Bible subconsciously into a "fairy tale," yet they dumbed down people because we all were taught a fairy tale!

I wholeheartedly believe that it's a lot worse having your mind imprisoned than having your flesh/body-self because if your mind isn't part of truth as it actually is and you were taught opposite of what truth is, whether it's factually or spiritually true, then the only outcome becomes distanced from not only truth but also from God as well! As kindergarteners, we were poisoned from innocence of any cognitive knowledge of any truth at all, mostly experimenting as our young brains try to develop its sea legs! This transformation of pure thought to expanding your very ability to think for yourself both metaphorically and metaphysically too. The spiritual realm of thought has been suffocated as a natural bridge to God, truth, and ever-expanding knowledge for oneself. The only thing that bothers me about everything is that our children continue to be taught the very same lies of theoretical physics as I was, and elementary schools seem to be the only choice for Buzz Aldrin to visit nowadays. The lie of being the second man on the Moon has created more alcoholics than any profession, and this is really sad. I do not believe these ex-astro-nots will be visiting any adults from most anywhere today that can think for themselves instead of being told how to think, what to believe, and what not to believe because, just as they clouded our blue skies with their poisonous chem or comtrails, there have captured most young minds and have thrown you in prison. The sleeping sheep are waking up at a faster rate than any time ever in modern human history, and the necessity to expose all to the truth is imminent.

We all need to take our minds back, for they are in prison, for we have gone ever so deeper down the wrong to zombie-like ignorance, and it scares the heck out of us who are simply observing the world of utter hate laced with conclusive incognitive behavioral thought. There's no single human being that I would challenge in any debate regarding the flat Earth-geocentric biblical in Scripture, as well as common sense against the theoretical psychics that is totally fabricated and illogical based science using math equations to try to illustrate an unprovable point that has manifested the imaginary to our real-belief system of reality. The science of psychics came from alchemy, which is black magic and the occult. It has manifested the biggest lie from the very devils of our society that holds the very highest positions of power and wealth. I will also include some Scriptures that our creator undoubtedly referred to the very shape of our world is the direct opposite of most people's belief system. God most certainly said that the Earth is "immovable," which is exactly what your very own senses tell you. You and I have never moved one inch since we've been born. It's very simple to rely on what God gave all of us, and though some are very challenged to be able to connect common sense and the astute expandable awareness that is imprisoned in your very mind, the capacity to start asking questions is part of all. As long as people continue to believe almost everything the idiot box says and the "impossiball" that has manifested beast on top of other beasts with deception based on a ridiculous theory, then people are seeing the days of judgement that is ready to bite everybody's proverbial backside.

Some Scriptures that prove the flat Earth truth can be found in Job 38:14: "It takes on form like clay under a seal, and stands out like a garment." The historical record of having important documents notarized was done with either ink or clay. The seal presses down on the clay that was called a segment ring and creates a dinner-plate-shaped circular seal with up-turned edges, just as an old official clay seal would look like, and our creator was describing the shape of our world. According to Job 38:4 "Where were you when I laid down the foundations of the Earth?" In Job 26:10, He drew a circular horizon on the face of the waters. Wow, our creator described the "ice wall" circling the waters of our Earth world. Would I go even deeper to suggest that Antarctica holds the secrets to the very existence of our creator? The glass plasma dome firmament was discovered in 1956 by Admiral Byrd, and the country that nobody owns was quickly annexed for as long as the parasites who run our world are in power. Isaiah 66:1, thus says the Lord, "Heaven is my throne, and earth is my footstool. Where is the house that you will build me? And where is the place of my rest?" I believe in Daniel 4:11 is pretty telling: "The tree grew and became strong; Its height reached to the heavens, And it could be seen to the ends of all the earth."

You cannot take this literally like the asleep are doing, thinking what is written is practically just a fairy tale because we were taught gradually in school to poison our mind from all good and what is powerfully joyous when you connect your spirit to your flesh because modern man is divorced from the truth, and that truth is at the foot of God who created all of us. In Revelation 20:7–9, Now when the thousand years have expired, Satan will be released from his prison and will go out to deceive the nations of the world (fake curvature camera lens-Convex/Concave" fish-eyed lens, and fake-space, lies of theoretical psychics and scientism, Darwinism, i.e., we all came from monkeys, Giant Nephilim, Evolution, Dinosaurs, Global warming,) which are in the four corners of the world. Take careful note that a spinning ball of any kind certainly doesn't have four corners. To continue in verse 8, Gog and Magog, to gather them into battle whose number is as the sand of the sea. They went up the breadth (means *pla toos*, an old Greek word *platos*, plane level and flat) of the earth and surrounded the saints and the beloved city. The fire came down from God and devoured them. In describing the protection of the religion called Islam, it's described in Revelation 6:7–8: "When he opened the fourth seal, I heard the voice of the fourth living creature saying, 'Come and see.' So I looked, and behold, a pale horse. And the name of him who sat on it was Death, and Hades followed with him. And power was given to them over of the earth, to kill with sword, with hunger, with death, and by the beasts of the earth." I do not dislike or hate any human being, and I do respect all cultures of all peoples of the world. Most of the Roman Catholic and Islam followers and the Hindu and Judaism faiths just don't know what is the transparent meaning in their so-called holy books nor what is not. By having a Bible, Torah, or the Koran on your night table collecting dust is absent of any knowledge that's inside the book, and very few experts will emerge on the very meaning of the actual content. I was raised a Catholic and attended schools in a heavily Catholic city run by the archdiocese. I can remember when Pope Paul came to New Orleans thirty years ago, and the locals were beaming and seemed overbearing with seeing him or wanting to see him It seemed a little repugnant in the similarities that some of our delusional folks and their narratives were looking at him as a savior or what the reaction of what it would be like if Christ Jesus was visiting. All religions were created a long time ago by these parasites at the top who hold over 98 percent of all the wealth and literally own almost everything and will disavow what they don't own. All five major religions were established by these children of the "watchers" for these religions to be at odds with one another and to fervent anger and then death by war or other means. We have over 7 billion people on the plane, and we also have 1.4 billion Muslims who lost their land of Palestine in 1948, when the state of Israel was established mainly funded by the Rothschild family. It is biblical in content that God said no man can provide any land to the Jews, and only he can provide a sanctuary called Israel when they decide to accept him and his son Jesus as the savior of man and mankind.

It is utterly mind-blowing to me that the Jews and the Palestinians have been fighting for over two thousand years, and that should confirm any notion that man created religion to fight and hate one another. Some other Scriptures to prove the flat Earth truth are many, and that number does exceed over seventy-five. One of my friends from over a year ago tried to present me his own biblical proof of a ball, actually trying to prove the Earth is round. He used Isaiah 40:22: "It is He who sits above the circle of the earth, and its inhabitants are like grasshoppers, who stretches out the heavens like a curtain, and spreads them out like a tent to dwell in."

This verse actually does suggest that the Earth is round, and I fully agree with what our creator states. However, a pancake is round also, but it's flat as shit, isn't it? A coffee table is round but is also flat! Interestingly enough, our Lord went on and suggested the firmament, being a protection as a stretched-out curtain like a tent for all of us to dwell in. This describes a reality of where we live, and though I didn't take my friend on personally with any rebuttal only to save any argument as to turn the other cheek, my answer to him would have probably caused some divisiveness between us. He has been going through a lot of marital and financial problems, and I simply didn't want to fan the flames of his depressed state.

THE FINAL THOUGHTS

These are scary times to be living in, and as Hurricane Harvey ripped into southeast Texas near Houston and wrought the same type of devastation that twelve years earlier almost to the day that Katrina destroyed most of New Orleans, including my home in Mandeville, we give our sincerest thoughts and prayers to all who are affected by this storm that dumped over four feet of water in many places such as Dayton and annihilated Rockport, which is completely destroyed. It is still too early to tell if Texas storm casualties will reach the 1,800 loss of lives that it did to the Crescent City, but it's quite obvious that it's still too early to ascertain a final number in Houston. The advent of our super technology of cellular and the internet has most likely enabled the beginning of the times of "tribulation," and the evidence is so clearly right in front of all who can see.

The eyes are useless when the mind is blind!

Research Flat Earth

We are living in the most perilous of times where Americans are attacking other Americans because of who they voted for in November 2016 elections, and the garbage truck called CNN is instigating civil unrest. They aren't the only ones, but they seem to get a lot of the attention because of their inability to link the KKK of 2017 left-wing radical communists thugs or middle-class white rich kids who are apologists and play a redeeming role for the deception of hating their own skin color while coddling to the rest of a group called ANTIFA (Anti-Fascists), who are actually very much a fascist group who are getting up to $25 an hour from the George Soros groups. Soros is a Hungarian American multibillionaire who is the Albert Pike of 2017. He is eighty-seven years old, and his fortune exceeds $35 billion. Forbes Magazine has him listed as the thirty-third richest man in the United States. His paid hate groups seems to be a key for a "global currency" by any necessary means possible, and being a lieutenant for the Rothschild/Rockefeller alliance bringing about the total destruction of the United States of America, and they have already been overwhelmingly successful in disconnecting man's conscious state from God and his true importance of his true destiny and the fact that we were then and still are made in the very image of your creator. In no way are we part of an infinite cosmos with nothing to connect your conscious state of being significant. On the contrary, we were taught that we came from apes! The failure of education and our true manifest destiny of humanity and what it means to be geocentric and not heliocentric has been a burdensome reality for myself, and I am not regretful to run away from the truth. I certainly cannot unsee the truth and will never go back to being in an imaginary delusional state of the Star Wars belief system.

The very subject of what these extraterrestrials really are is contrary to the cosmology belief of being in a static state of being in a land of make-believe. I recently shared my thoughts to a panel discussion on Facebook about the immense repercussions of dumbing-down humanity through the lies of education. I responded to a friend of a friend who was lambasting my friend for being spiritually connected with Jesus Christ and is an A-plus conspiracy theorist, especially with the 9-11 inconsistent entanglements that allowed several canisters of worms to be exposed that definitely draw a red flag. How in the world could anyone make a cellular phone call to their loved ones at thirty-five thousand feet in the air? It is quite impossible to wrap my head around this because the triangulation of using three cell towers to accept any call is highly incredulous to believe because at ten thousand feet is the maximum height as advertised. Even at ten thousand feet is quite impossible, so a lot of what happened during the hijacked flights who allegedly made calls to their family members is not possible.

In response to my friend being dissed by one of his workers who was mocking Jesus or any proof of the existence of Jesus Christ was followed by my reply to my friend Bryan.

> That's the great education, I mean, the poisonous indoctrination of the public school system that has effectively helped create humanity into a "race of idiots," and it disconnected most from real truth and real knowledge. This is why most people cannot connect to their spirit because they are accustomed to being told what to think, when to think, how to think. Arguing with someone like that isn't worth it. Lots of people are stuck on stupid. Just pray for him.
> —Mark Steven Hollander

We have a definitive divide in the United States today, and the global elites are absolutely beside themselves that Hillary Clinton wasn't elected last November 2016. Even with the enormous wealth derived from the George Soros-backed Clinton Foundation and the enormity of the upset that found Freemason Donald Trump beating his fourteenth cousin by marriage, Hillary Clinton, in the most soap-opera filled and the most fascinating election in U.S. history. In my humble opinion, this far exceeds even the George W. Bush-Al Gore 2000 election. That was probably a prelude to this one, notwithstanding the "hanging chads" that created a scenario that seemed like Ripley's Believe It or Not. The hate in the United States today is like nothing that I have ever seen. With the disconnection of truth resonating everywhere, from the mainstream media and the claims by President Trump of the media of creating and putting out fake news stories and the repulsing narrative of this president being literally attacked at every word that spills out of his pie hole. The Hollywood weirdos have come out of their mansions and going political and leaving their land of make-believe characters who are only relevant for our entertainment. Since February of 2016, the box office receipts are down by more than 26 percent, and it still suggested that these neophyte, liberal, from the land of the imaginary world—thirty-third Freemason—soulless and clueless Hollywood elitist are trying to get the president impeached for the sole reason of not accepting the outcome of the election from more than eight months ago. I do not know if Trump will be able to ever do the job, as the left-wing wackos have come out of their closets and will not let this man do his job. It is absolutely mind-boggling that multibillionaire George Soros, I Globalist, who is trying to destroy the United States and helps create the CNN's of the Jewish banking-owned mainstream media networks very narrative of reporting artificial and phony news in order to create complete and total chaos in the streets of America. Back in 1984, when Pres. Ronald Reagan was the commander in chief, he suggested in a speech that he gave, "If FASCISM ever comes to America, it will come in the name of LIBERALISM."

Well, it's here, and they dress in black and cover their faces with black bananas and are today's KKK and are financed by the Soros groups with full intention to start a civil war between the left wing Democrats and the right-wing Republicans with the ANTIFA group, which the FBI has deemed an official terrorist group and aren't what their name suggests. These communistic fascists are bent on destruction of human values, from monuments to Trump to absolutely destroy anything getting in their way. The Charlottesville caper was very tragic in that one woman died from the violence that was reported by all the mainstream media as being started and propelled only by the KKK, neo-Nazis, and the right-wing hate groups opposing peaceful opposition protesters. In the world of lies and the reluctance from the truth, the media failed to report that the ANTIFA and the BLM (Black Lives Matter), Soros-financed groups, are bent on its hatred of anything American. The lost is going to eventually speed up exactly what the elites are shooting for, which is to have one world government with no borders and divided into thirty-three sections. The monuments in New Orleans were removed over a four-week period, and the city spent $2.3 million taking them down. Mayor Mitch Landrieu was definitely paid a whopping amount of money to bring the ultimate divisiveness that created a scar of distrust and hate in the New Orleans area. The city is completely broke because of the massive corruption of people of Landrieu's girth, and while pulling the New Orleans Police Department officers away from their regular patrols of my beloved city, sixteen people were murdered to protect the protesters from the right and the mostly ANTIFA protesters from the left, which had been flown in from California and New York by the Soros backed protesters and were paid more than $25 an hour to cause harm to anyone who was a Trump supporter, or the attacks from waving an American flag and being accused of being in the Klan (KKK) and sometimes were beaten merely because they are flying 'old glory'. It amazes me that you can clearly clarify the difference between the hate groups who propagate the hate just by who exactly comprises their makeup. I cannot resolve the fact that the CNN host Kamau Bell was speaking at an ANTIFA rally near the University of California-Berkeley and was encouraging hate like I've never seen from a supposedly objective journalist and anchorman from one of the major networks in the world.

The outrageous time of insanity are definitely here, and if this man doesn't get fired, then it just proves that President Trump is on point and correct about his accusations that most news networks are reporting inaccurate news and are lying to the American people each day. His spewing hate not only help incite a riot, but one person holding a camera was beaten almost to death. Fourteen of these misfit thugs who are no better than that of human feces were arrested. Their mugshots tell the story of their backgrounds, a few uppity class whites who can't stand themselves for being white, meth and crackheads who no one would ever dare breed with, and common criminals who are either programmed by the MK-Ultra program by our government to wreak havoc and create civil unrest and outright war. They are thugs of society, but many of them are beyond comprehension on just how illiterately their thoughts are seems to be focused on destruction and hate for President Trump. I will not let the KKK and the neo-Nazi alt-right groups off the hook because it takes two to battle, and the propagation of hate or from the money controllers who are trying to bring in the new world order without any borders. The Ku Klux Klan who, by the way, was started by the Democratic party; but time and time again, and I will preach this same message that humanity has been so lost in their sense of self-worthiness. No spirituality for any deity at all. In fact, we are producing the lost atheists at an enormous rate. Most of humanity cannot think past the bridges of their proverbial noses. All hate are to be condemned, and it runs the opposite of any intellectual forthright of common sense. Today people believe anything that is so ridiculous, and the gullibility of people have convinced me that the satanic apex point of dumbing down humanity has worked successfully. The fight is on for me to take as many wools over as many eyes as I can and to always be truthful and live and die with the knowledge that my creator has so empowered me with. I know I am obligated to share it with my loved ones and while I'm on tour speaking to the curious masses.

The following Bible verses will be confirming the true destiny of man's potential both mentally as well as physically. Let's go right to the closest understanding of why our world was twisted into a false doctrine that perverted not just what was taught but was also allowed to be taught and almost exclusively unchallenged or silenced through payoffs and bought off to continue the ridiculous theoretical teachings without any evidence to support said doctrine of thought. My sources of being peaceful is that my best friend is Jesus Christ and that I talk to him and lean on him all the time. I trust and have faith that everything is done in his time, and it has increased my faith to know that he is a merciful God, and his grace is beholden to us all. It is so vibrant and so fulfilling to experience joy that I get each day, and it's a joy that is indescribable to anything that this world could ever offer. Yes, the world's temptations are many, and a lot of what is fun in this world is exactly just fun. Joy, and I mean real joy, comes only from God. Let no man intercede between you and your personal relationship with the living spirit of God, simply nothing like it. Another inspiring way for me to have complete peace and absolute joy is to try to get in a mood to transfer my research and the years of being able to accumulate knowledge that has even been a little overwhelming at times. I listen to *Phantom of the Opera* soundtrack, and it is the music that soothes my soul and places me in an even keel with being completely unbothered by anything else. Probably my favorite music would be of Pink Floyd, who wrote extensively about "flat Earth," and it is throughout the whole chronology of their music for over forty years. The song that really defines my journey into the truth would be a song that puts me into a writing mood of continuing my quest to complete this work.

Pink Floyd, "Learning to Fly":

> Into the distance a ribbon of black stretched to the point of no turning back. A flight of fancy on a windswept field. Standing alone my senses are reeled. A fatal attraction is holding me fast. How can I escape this irresistible grasp? Can't keep my eyes from the circling skies. Tongue-tied and twisted. Just an earthbound misfit!
>
> Ice is forming on the tips of my wings. Unheeded warnings I thought I thought of everything. No navigator to find my way home. Unladened, empty, and turned to stone. A soul in tension, that's learning to fly condition grounded but determined to try. Can't keep my eyes from the circling skies. Tongue-tied and twisted. Just an earthbound misfit! Above the planet on a wing and a prayer, my grubby halo, a vapor trail in the empty air across the clouds I see my shadows fly out of the corner of my watering eye. A dream unthreatened by the morning light. Could blow this soul right through the roof of the night. There's no sensation to compare with this, suspended animation, a state of bliss. Can't keep my mind from the circling skies. Tongue-tied and twisted. Just an earthbound misfit!

The Freemasons have been ruling the roost with much of the world's wealth for over one thousand y------

One thing that needs to be perfectly clear, and I mean this with all of my heart and soul, is that the lower-levels of Freemasonry is not the subject of my scorn. Rather, the highest levels of the secretive society is the revelation that the Catholic church is but a deceptive cloak or an artificial shield to hide the real intentions and the only agenda that has most of the funding form the world's resource of wealth. My father's best friend was initiated into Freemasonry but never really stayed in it, and it was probably problematic only because of his inability at that time in his life to be able to attend a regular regimen of its activity schedule and they have and done a lot for humanity causes and charities all over the world. So this isn't about the ones who realize not! It most certainly is about the ones who cover the light of purity but the ones who turn off the light of good and sell their righteousness nature for a table of wealth and fame!

I don't pretend to have all the answers concerning why 2017 could very well be one of the global elites push to unleash the same nonsense with CERN and the halogen collider in Switzerland, by colliding atomic particles to produce portals to unleash pure evil to wake up the inter-dimensional spirits in a modernistic way but with the same results that the twelve ancient tribes of Israel did by worshipping Baal. Baal was a demigod and one of the pagan Sun-gods that influenced the Jewish people to completely turn away from God and became the original ideology of the demigods who fell from God's grace after the war that evidently took place in heaven just over 4,500 years ago. Baal was an imposing figure who almost had the face of a cow with horns and was a principal of hell. Several maids of Israel would sacrifice children to the altar of Baal. The ancient tribes of Israel had persecuted and perverted the idea of what is normal and what isn't. His stance has been emanated throughout the Freemasonry ascension into darkness, but they somehow call it "into the light." What a croc of shit! I will say literally up was down, and evil was good, and good was evil. Christians were killed and tortured to a point where we must draw a correlation with no absolving innocents can only be one's conclusion that leading into the days of Noah, especially during Jared's reign, the children of the pagan gods, who are better referred to now as bluebloods.

The population of our world in the last decade has given statistics that only 12 percent of the population are sequenced with DNA that has other characteristics. One that would include the Royal family in Great

Britain and the Queen Mum herself that when Prince Charles chose Lady Diana Spencer for his wife, the Spencer family were in fact blue bloods. In no way, shape, or form can any royal blood or any blue blood, the global elites, etc., marry or convalesce with the other 88 percent of the world's population because more likely the newborn will be deformed. I will not get into this subject at this juncture but will entail why this is the case and why the anomaly occurs.

The thirteen ruling families of the Tri-lateral Commission IMF, the Club of Rome, and certainly at the top would be the Rothschild family-owned World Central Bank. We are taught that the Civil War in the United States was because of the slavery issue, and of course, it was a horrible injustice that was propagated against African Americans and the slave trade that was a huge profitable venture by the very rich elites, and almost all are of Jewish descent. The Civil War was a gross injustice done to our fellow black Americans, and the hypocrisy initially that we do know that the North was also guilty of harboring slaves from West Africa and leading into 1861, the George Soros of the day, Albert Pike, a Freemason whose main attribute was to cause complete chaos in America, and that he most certainly was almost like a double agent. He was sent by the controllers of the World Central Bank to instigate a war between the states because President Lincoln and the United States Federal Reserve Bank were never going to cede America's money to a foreign entity, and in this case, the war was actually fought to see who would control America's money as well as future revenue concerning how it was controlled. The very reason Lincoln was assassinated by another Freemason, John Wilkes Booth, because of the president's unwillingness to sell out America. Because the Civil War was not about slavery, it was a humanitarian disaster where the backdrop of such a hideous treatment of our fellow citizens was exposed, and the mere fact that Albert Pike was sent to not only infiltrate the very inner linings of how the United States was going to grow agriculturally, politically, economically, and the powers that have ruled the roost because of the enormous wealth that they have accumulated going as far back as the crusades, the very bloody battles between the Persians and the Crusaders. War and fear and killing one another normally are at the top of the agenda for the controlling European bankers and their allies who are definitely heirs of the pagan Sun-god worshipers. The Rothschild and the controlling power of what money and enormous wealth can do and what it can create and cause. The one common denominator that is especially alarming is every single war that has been fought with the European bank as the main financier for all armies, friend, and foe. By golly, they financed Adolf Hitler's Nazi war machine. The U.S. war machine, in fact, President George Herbert Walker Bush's father, Prescott Bush, was also involved in very nefarious dealings concerning the very bad deeds of the Rothschild's and the controlling factor that all the world's armies during wartime are now and have been financed by the Rothschild-controlled banking system, and the fact when God flooded the world to try to wipe out these demigods and giants who depopulated the world because of war and sacrificial murder of children, the evil that was the days of Noah. Obviously, the intent was to destroy the evil of the pagan gods who procreated with woman and the offspring that were called the Nephilim. Eventually, the ancient pharaohs of Egypt were mostly gone, but the ones who did survive eventually found the new haven—the United Kingdom and certain areas of western Europe. Of course, Buckingham Palace was a prelude to form another country called the new world to instigate another war to accumulate the already-massive wealth that they always used to control the masses. I am not speaking at the local level. The intent was to control the world and to make sure that the numerology, which is very important to them, to try to speed up the process to form the place of the mystery Babylon or the New Babylon. The Scriptures in Revelations are especially spooky as far as who this country is, and the similarities of the description is eerily right on target, describing the United States of America.

I cannot see any intervention that a common man could do against this propaganda machine that has not done enough to serve the betterment of mankind and apparently seeks to deliver a blow to everyone with a type of manufactured technology that is ready to wreak havoc on the East Coast of the United States but also the West Coast. The reference point of embarkation to try and eliminate 90 percent of the population of the world, yes, you did hear me right, the elimination of 90 percent so as to control 10 percent of the population that has toppled almost nine billion people. A document dated in July 2001 from NASA'S website outlined this and suggested that mankind is not worthy to be sanctioned as an important element in the nefarious plans to bring about the cyborgs, the robots that are now being developed as a fighting force not to protect us from evil intent but to propagate a dilemma that will manifest and speed up the times of judgement. I'm talking about Jesus Christ coming back to save humanity from the deathly deception and the employed motivation to destroy man's soul. The consciousness of new information has been completely in place to dumb down all of our senses, our brain's ability to receive new signals of information and discern proper critical thinking. The exact feeling that my spirit revealed to me is that the chemical trails, the Freon repellants, and the dosage of sodium fluoride all over the world will not stop. In the New Orleans area, it's every single day, and I have to surmise that its' hard to even suggest this to my wife, Alicia, or any of my loved ones—my three daughters—because I actually know that every time that we brush our teeth, we are being poisoned. The mere fact that everyone or most everyone has sense enough to realize that money is what drives everything and everyone on the face of our Earth.

The twelve ancient tribes of Israel, even in the Old Testament, defied God virtually at every step of the way, and the powers who not only rule the global establishment but also the financial institutions, including all the stock markets, all of the schools of the world, and the textbooks given out to small children who we are all innocent, especially at age five and we are very nervously not knowing what to expect with our developing minds, and the growing development of becoming a tool for the school teacher who also knows only what she had been taught, and accepted it without question as being a given truth. Unfortunately, the so-called educators, especially our esteemed college professors, the astrophysicists department heads, and the whole basis of the fraudulent cosmology of an infinite universe, was built on a model of backward mathematical equations, dark matter, and black holes in an unending and ever-expanding universe that just perpetuates one lie upon another. It seems to me on a bright beautiful Thursday in March 2017 that there doesn't exist any scientist or physicist that can explain how the ever-magical laws of gravity works, especially at the experimental stage. The law of gravity was only enacted into law by people who have a very nefarious agenda, and it goes without saying which rings true, "The UNSEEN RULE THE SEEN." The Freemasons cannot do any of their dark plans without numerology, and the significance of this mostly unnoticed deception that is clearly in place before any destructive event that would take place to continue to defame humanity, to deny the truth to humanity, and to deny the existence of a God, only pagan Sun-gods that they continue to worship, especially Baphomet, who in essence, is Satan, or Lucifer himself.

We will cover one of my favorite proofs of what our reality really is, and the reasons that I will document will be of an unquestionable proof that we are in a geocentric and not a heliocentric reality and that it is exceptionally very difficult for people of modest educational means to accept this notion that they have been taught a bunch of lies through education and the main tools of the global elite control system, the news media. They control all commerce in every country except for North Korea, Iran, and I believe Freemasonry will be in Cuba since the recently lifted trade embargo was lifted by the Obama administration. The occultist and the satanist Aleister Crowley was a main influence in being able to summon spirits to inject into the acting abilities of portraying the character that scientologist Tom Cruise and most of all of Hollyweird induces into the psyche of the Freemasons willing to move up the ladder in rank, all the way up before the meeting with one of the head wizards, the Grand Sovereign Commander, in order to eventually sell your soul from a contractual and business acumen—to Lucifer. This enables the continuation of being a big star and, in many cases, the ascension to superstar status and the Hollywood indoctrinated poison that copulates a different reality that forces a different realism and clouds your moral compass when you are unable to figure what is wrong from right and to suggest that up is down and down is up and forces guilt upon a good moral person, who, after continued poison after poisonous and venomous fantasy about things like our Star Wars space adventures reality prism of so-called truth. According to them, what it actually does is condone you and your reality and effectively prevent any new information to penetrate your brain, your senses, or anything to do with anything that would not line up with your brainwashed and forcefully docile mind is quickly spun out of your line of belief and then ridiculed and insulted with vulgarity and then made fun of. You may think, "How dare anybody who questions the great scientists and brilliant alchemist, occultist who practiced black magic, and his name was Copernicus, an absolute thirty-third Freemason who was so terrified to even publish his works of a spinning basketball Earth that his theoretical thesis and his complete works of Earth spinning around the Sun makes absolute perfect sense, to ram this down humanity's throat, as a means to hoodwink humanity into becoming exactly as they are, to sell your soul to the devil, to renounce Jesus Christ and the truth that was always told in the old Hebrew Text of the Bible, and the last five hundred years we have had the greatest lie in the history of humankind, and this is the sole reason that Copernicus was forced to publish his manifesto from the Jesuits who are well funded by the thirteen Jewish Families. The Bilderbergers, Rothschilds, the Queen of England, and Buckingham Palace, are all absolute subjects to do Satan's work here, and the blindfolds have poisoned almost every one of the truth. Just as the Bible says, their time is short, but they are pagan Sun-god worshipers and had to produce the main model of this worldwide deception, for Lucifer is the master of deception.

If you even question what you have been taught to not be true is very difficult and, in some cases, impossible for the human mind to expand cognitively and to even try to look at the other side just to ensure yourself that you can ascertain and keep this thought as knowledge to know that what you learn, or unlearned, to be an absolute truth, not just foolhardy theories (evolution, heliocentric, dinosaurs, Nephilim giants). We are going to see at least a chronology of how the Templars became the Freemasons of the thirty-third order, which is the dark side of the ascension called "into the light."

A quick look at the very history of the Templars reveals a major progression and transformation they underwent along the way. They first appeared under a Christian façade, but soon afterward, they entered a darker phase in which non-Christian and perverse philosophies and teachings show through and became transparent to the changing events that are solely responsible for those changes. When it initially came about, it was during the Knights Templars' sojourn in the Holy Land. That's when the Templars became acquainted with the Cabala and learned all about the different practices of the various Jewish sects. The assassins' mysticism was put into great use during the time of Martin Luther in England, and the war against Christians and the Bible's word with the Protestants became all too real. I can remember celebrating St. Patrick as a true symbol of a good Catholic's yearly ritual of this Holiday that is celebrated in New Orleans with a parade and revelers getting drunk and giving out cabbage for an exchange for a kiss. The riders in the parades were mostly men, and thus it was the women attendees that usually made out with the most cabbages once the parade was concluded. The war with the Irish, who, in some cases, were burned at the stake alive with the Holy Bible around their necks were so horrendous that it's hard to comprehend this to be ascertained as a fact. However, numerous accounts have verified that this was in fact the case. Before the time of Martin Luther and the old adage that a tremendous amount of money was part of the partnership with the main Jewish sects who did control even then, most of the resources of money and the highest orders of the IHS/Jesuits and the Jewish congregation of ancient banking acumen was firmly in place. The trading of the Christian faith had given way to secret "occultist rituals" and "black magic rites," which changed the ideology of most Templars from the highest ranks who kept the secret highest agenda away from the innocent Templars who held and kept the faith of the Roman Christian Church in Rome. Again, the major transformative cause of the change of the Knights Templar's Order is when they acquired incredible wealth over a very short period, with the promise of attaining mystical powers over the material world through their ritual practices and dark beliefs. They then set their sights on a much grander scale, especially at that time, unlike the super-informational highway age, the higher Templars didn't reveal the mystical and secretive beliefs of acquiring great wealth with the teachings of seeking help from the dark powers of black magic, and another facade was that they used scientific methods, just as NASA is using in the twenty-first century to continue the heliocentric lie in order to separate man's true so spirituality for his creator, and the convolution was to try to contact the invisible powers by means of secret codes and numerology. They also used magic signs (33, 322, 666) and formulas (Albert Einstein's theory of relativity) as well as incantations. Poisons were prepared, and the elixir of life was sought in these scientific experiments or an alchemy confirmation. The attempt by alchemists to make gold out of lesser medals convinced the Templar's of the highest order that the marriage between the Jewish elitists would together rule the whole world, and the resources were solidified at this time, just as it continues today. The ultimate sin against God is that they chose to worship Satan and called on him to dominate the powers of darkness, thinking it will keep them out of the dark realm. The investigations by the courts of the King and the Pope documented the Templars' nefarious ideals, and it exposed them to indicate that they were using their Christian faith as an affront to gain favor with the unknown masses who had absolutely no knowledge of what was taking place at the highest levels at the Vatican. The Templars mass influence combined with the Jewish money created a union of deceit that still continues in 2017. The dark symbols also founded the new acquired wealth to build the very castles that these powerful corrupt leaders of yesteryear and are very much put on a pedestal and traditions to rule over their people who essentially started worshipping them instead of God and Christ Jesus. This led the way that amplified the eventual Freemason secret society. The Templars infiltrated many different sects and organizations that later created all the Masonic temples and lodges. Eventually, the Templars brought the different cultures in most of the world's peoples under their control and coaxed the populace to accept the philosophies, beliefs, and rituals. The Templars were trained in the arts of architecture and masonry and the many European Castle and the beautiful Gothic Cathedrals of stone masonry which was an addition to the symbolic features of the mergers between the Masons and the Templars. The symbolism is so nefariously but brilliantly created, and the deception is so strong and that when Eve bit the apple in the garden of Eden to tempt and later defile mankind's true nature of God's creation, it was replayed through a universally and monumental deception from the year 1666 with the thirty-third Freemason Sir Isaac Newton playing a role in trying to prove gravity with his alchemistical parallel of an apple falling from a tree and writing *Principia*, which is three times thicker than a Bible. Yet the word "if" is written in his magically deceptive book over 2,600 times! Why the irony with an apple tree? You guessed right by pillowing what was good in the garden of Eden with the duality of the satanic realism of tricking man and convincing all humanity with a fictional force of "the gravity God" of false realism.

Another union of the Masons and the Templars is the grand master's abacus (staff of office) in determining the symbol representing Aaron's rod, which in the Bible means a walking stick that sprouted leaves and a head forming into a temple and along length of its body is carved measurements. The staff is the symbol of masonry. It is esoteric in the nature, and the architecture is the Gothic style that you see throughout Europe and built deliberately in the new Gothic and copied after Jerusalem's conquest by the crusaders. With the Templars' grand master being also the Freemasons', the Cistercian monks, dealing with construction planning, had also been members of masons' lodges, an example to the clerical or monastic type of mason. In Paris, the Grand Lodge for most of Europe and the world, also where the French Revolution started, essentially the starting gate for internal strife and the war to overthrow the French monarchy, is the place where in the eighteenth century, the masons and the Templars shared the same lodge, which accents the close relationship between both parties. The papal decree of 1312 that liquidated the Templars order also ended the Masons' right of free passage. Fearing even worse reprisals, the French Masons fled to Germany where Gothic architecture dominated almost all construction. There, the Masons' lodges that received Templars' escaping from France experienced the same gradual transformation as the British ones had- from

1. from operational
2. to speculative Masonry.

The first handwritten Masonic document called "Regius" was adopted in 1390, when the lodges were indoctrinated the ideas of the Mason's with the apparent façade of using the Vatican because of the sheer amount of the world's population, and also the very doctrine of ordaining Peter the apostle as the first Pope, or Pontiff. The age of human history must have other ancient knowledge concerning the Mason's grip of most of human resources that are bartered or the exchange of goods, which today amounts to the money that is printed. It is said since the start of modern society, that power is money, and the use of the powerful and corruptible resources of Christianity and the resources together has formulated a tandem that is infested with a very nefarious agenda. Another important aspect to all this has to be that the written word can and would be leaked out, and the secrecy of the order of rules within the secret organization was better left without written documentation before 1390. The Masons survived with their secrets, safe within this order, that is until the Templars were annihilated and abolished by the Inquisition. It was then when it was revealed that the Templars and the Masons were of the same rules. For the two hundred years, the Masons and the Templars shared quarters. It's quite obvious that they influenced each other, and in fact, the Masonic rituals seemed to have been copied from the rituals of the Templars.

The link between these two-headed beasts are from 1760 that makes no secret of their shared inclination for mystical knowledge. The document was handed down from Jacques de Molay to contemporary Freemasonry, and the German Rosicrucians is unquestionable. The Templars infiltrated the lodges of the weak and passive Masons with the organizational skills that the Rosicrucians would develop into a strengthened order and was a tremendous tool. Templars are not a branch or aspect of Masonry, but Masonry, with all its symbols, ideology, numerology, and history, has become a den or pushover for the Templars, obviously under a different name. Masonry is linked as far back from the Temple of Solomon, and the mystic symbols came from the Cabbala, but they adopted their organizational structure from the Templars. In addition, their ceremonies, oath, dress, and rules of promotion being prepared according to the Templars rule all prove that the Templars and Freemasons are one and the same.

The Scottish Rite of Freemasonry is the oldest Masonic lodges employed to provide shelter for the Templars and is dated back in the fourteenth century. After the Inquisition, many of the Templars fled to Scotland.

The Scottish rite of Freemasonry was during the eighteenth century and started by Baron Karl von Hund. His fame in the order is legendary, and when the Templars fled from the Inquisition, they were in Ireland. The reorganized secret order became a stronghold throughout Scotland. Von Hund created the Rite of Strict Observance, which eventually spread throughout western and eastern Europe as far away as Russia. It's little wonder that the Masons control not just NASA, but they are one narrative, and they are reading from the same green screens "of all the world's space agencies."

It is of little doubt that these guidelines were the modern orders only able to reveal to the very few. It also created the order of the Temple. Which is the highest of the chivalric orders in the York Rite, which formulated the thirty-second Scottish Rite. No question that the blueprint for the thirty-third Freemasons and the Illuminati have concurrency of its origins to Baron Karl von Hund!

The writings also gave a clearer interpretation of the tolerance and freedom of thought within the Religious, political, and the ideological realm of all society simply because they can create any type of fundamentalism truth or what they want truth to even look like, but the endless about of money most certainly help create the different false reality that we are living today. The Freemasonry, Templars, and the Rosicrucians philosophy of free thought were integral tools to bring a level of cooperative engagement with one another, to create the best picture of its thesis of the full agenda. The Rosicrucians was founded by the Templars and was considered a sister organization to Freemasonry but were the "dark side" of the Order, and I keep summarizing in my thoughts that the Rothschild family fortune is the origins of this German-Swiss-Dutch agenda driven part of its development and is what we essentially still see today. The Rosicrucians centers were an obvious place for the Templars to obtain the magical powers required to control the material world. In essence, these were the Nazi Scientists and the very brains of the order to summon the dark spirits from the dark side and also including the magical ability to open up inter-dimensional portals. We are experiencing the occultist inter-dimensionals appearing as grey aliens with these saucer-crafts literally appearing out of nowhere. The whole narrative of the lying global elite of today is to propagate this as UFOs from other solar systems, which is not only a lie, but it's a delusion as effective as would a hammer knock a human being senseless!

The other thing about the benefits that Freemasonry has done and continues to do is to promote wealth with even more wealth, but it certainly has had some positive impacts on the beautiful nature to ascend to some famous literature writers, actors, playwrights, sports heroes, and gladiators such as Alexander the Great, whose empire spread from Egypt to Russia before his apparent poisoning by a subversive operative because it was perceived that his immense power was completely out of control.

Some historians who are expects of the order fervently believe that Sir Francis Bacon was actually the true author of all the Shakespearean plays. He was a grand master of the English Templars and was also the most senior Rosicrucian. He was the undisputed expert in the secret sciences, especially the Cabbala, alchemy, and sorcery. In a more practical view to understand what I am reciting is that he was into black magic and calling out to every spirit of darkness and the practice of praying to any spirit except for Jesus Christ.

Bacon's *The New Atlantis*, which he wrote in 1626, is of a heaven on earth and an adaptation of the Templars' ideal state. He recounts the story of imaginary people living on an imaginary island called Bensalem, which means "New Jerusalem." The doctrine of the fable is an entirely scientific society, where the God of all is only technology. There is no creator, only a creator of materialism and of controlling the story or the narrative with the secrets that are well kept to the very few. This imaginary society is where the residents control everything—the people, the news, the weather manipulation—but even in 1626, he stated controlling the "winds." The false news on this story is that the real deeper, higher level of Freemasonry is controlled by the devil, albeit, the dark-side. The church is a phony facade and has used the cloak of deception to give all the appearance of Freemasonry being a doctrine of good and pure and backed by the world's largest congregation of organized religion, the Catholic Church. The narrative of truth that resonates its signature end is that "The wolf has come and is in sheep's clothing."

THE FLAT TRUTH: Your eyes and your reasoning as well as your own perspective—it was taught that what you see hear and what your instincts are telling you is wrong! You have officially been brainwashed to a different reality.

Angela Harris, Louisiana, USA

Tracy Cameron-Canada

Mark Hollander

Family, Thanksgiving 2017.

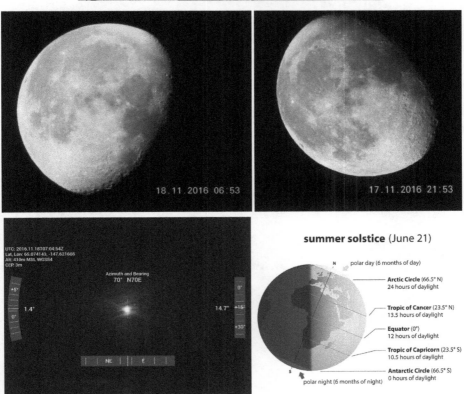

What my research has proven is by recording the speed of the moon, I have observed and documented the moon speeding up on its southern path towards apogee and on its northern path back to perigee. This completely destroys Kepler's first and second law of planetary motion.

Dave Marsh
Performed the Conklin/ Marsh experiment

Ex-West Point graduate, Shelley Renee - California, USA

Lightning Source UK Ltd.
Milton Keynes UK
UKHW011642030622
403946UK00008B/213/J